# PRAISE FOR *WINCE*

"There ought to be a warning sticker on this book—not about Wince's drugs and guns, not about his offensive thoughts about almost everything, not about the felonious and antisocial mayhem that he causes wherever he goes, but about the real likelihood that you'll somehow end up caring what happens to him. It takes guts to let a guy like Wince narrate your novel, but it takes a whole lot of skill to avoid caricature and to render up a real person. Smith's novel succeeds as a harrowing satire of American masculinity and violence, but it also succeeds, against all odds, as a complex character study and a surprisingly tender exploration of friendship and loyalty."

—Chris Bachelder, author of *The Throwback Special*

"Anton Chekhov warned us about the gun on the mantelpiece, but his characters weren't in the business of manufacturing weapons themselves. In telling the story of Winston 'Wince' Fisher, Jr., an irascible man with a penchant for AR-15s, Drew Nellins Smith gives the reader a front-row seat to one man's hunt for a better tomorrow—which, from this angle, looks not unlike a roadmap to what lies beyond rock bottom. Imagine a James Ellroy character in a Charles Portis novel—or maybe the reverse—and you'll have an idea of what to expect here."

—Tobias Carroll, author of *Ex-Members*

"*Wince* is a trenchant, timely and hilarious noir—imagine a Coen Brothers adaptation of a Ross MacDonald novel about contemporary McMansion Gun Culture, and that will give you an inkling of this sinister romp."

—Dan Choan, author of *One of Us*

"What will get to him first—a stroke or heart attack, the police, cartel enforcers, or his Uncle Gary? The well-named 'Wince' (short for Winston), a depressed, thrice-divorced, unreconstructed straight white Texan gun fanatic and cocaine addict who is running low on cash and obsessed with revenge on his old boss, is a guy whom you might not want to sit next to on the couch, but who is hard to take your eyes off. Drew Nellins Smith has created a singular protagonist who is very much of our times."

—Ted Conover, author of *Newjack*

"*Wince* is an audacious snapshot of our troubled post-COVID times—a world where you can order a gun as easily as takeout. Our guide through it, Winston Fisher, Jr., has no business winning our hearts, but he does. He'd be the first person I'd call if I was having a serious crisis. I loved this book."

—Patrick Hoffman, author of *Friends Helping Friends*

"Funny, profane, sly and bruising, *Wince* delivers criminal thrills with satirical wit. Winston Fisher is a hero (anti-hero, unhero) for our time."

—Sam Lipsyte, author of *The Ask*

"In *Wince*, Drew Nellins Smith has written a brutally funny novel about an American archetype we rarely see rendered this well: the aging gun guy. Winston is a ridiculous and detestable coke-fueled mess, and Smith narrates his commitment to self-destruction as a way of life so compellingly that it's impossible to look away. *Wince* shows that guns are never just guns—they're history, memory, manhood, and fulfillment of some of the darkest American fantasies."

—Dr. Andrew McKevitt, author of *Gun Country*

"Taking a smart, but arrogant gun nut/coked out slob and making him into a likeable character would seem like an almost impossible task for any writer, but that's exactly what Drew Nellins Smith has managed to do in his gritty, compulsively readable novel, *Wince*."

—Donald Ray Pollack, author of *The Devil All the Time*

"Starting over at fifty-seven is hard enough, harder still if you're a coke-addled criminal gun obsessive with a bum knee and an overbearing mother. Drew Nellins Smith's fast, funny, colorful, and surprisingly humane crime story is propelled by the strong, companionable voice of its narrator, 'the man, the myth, the legend, Winston Fisher, Jr.' a.k.a. Wince to his friends (and a growing number of enemies). Part joker, part smoker, part midnight toker, his life may be off the rails but he is a man redeemed by a code and you will surely root for him through one last, improbable score, with a bit of hard-earned self-knowledge as the ultimate prize."

—Mark Sarvas, author of *@UGMan*

"*Wince* gets at something about America's relationship with firearms with a vividness that policy arguments often miss."

—Robert Spitzer, author of *The Politics of Gun Control*

"Boldly bleak and brutal yet strafed with moments of screwy tenderness, at once a meticulous study in modern male dissipation and a no-bullshit Texas noir, WINCE is the aging-Gen X crime novel I didn't know I needed until I found myself unable to put it down."

—Justin Taylor, author of *Reboot*

"*Wince* is a Chekhovian dive into the spiritual desolation of the American West, where in the big sky ring the echoes of civilization's violence."

—Nico Walker, author of *Cherry*

"Wince is not just the nickname of the protagonist of this dark comedy, it's the reaction responsible gun owners will have to a character who uses his extensive knowledge of firearms to build and traffic automatic rifles to the cartel. But Wince is no cartoon villain. In the anti-heroic myth of Winston Fischer, Jr., Drew Nellins Smith puts a human face on the unsettling extreme of white male grievance culture, and that humanity is precisely what makes this book so disturbing and so hard to put down."

—Dr. David Yamane, author of *Gun Curious*

# WINCE

DREW
NELLINS
SMITH

ITNA

ITNA PRESS
Los Angeles, CA
itnapress.com

This is a work of fiction. Names, characters, places, and incidents are a product of the author's imagination. Locales and public names are sometimes used for atmospheric purposes. Any resemblance to actual people, living or dead, or to businesses, companies, events, institutions, or locales is completely coincidental.

Cover design by Chris Stoddard

Cover image "1957 Winchester 12 Pump Shotgun Ad." courtesy of Alamy



Drew Nellins Smith. -- 1st ed.
ISBN 979-8-9989551-2-9

Library of Congress Control Number: 2026937701

# WINCE

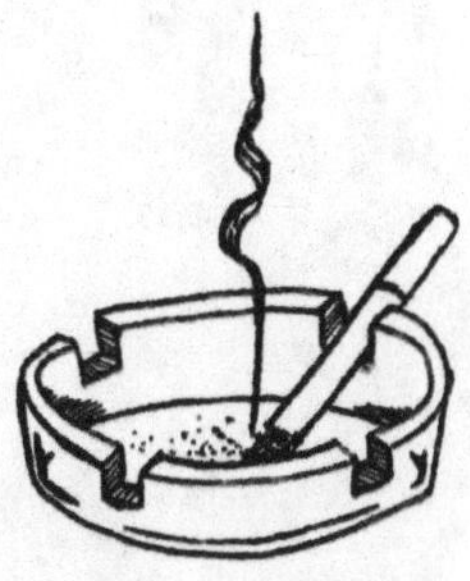

by

Drew Nellins Smith

# 1

LATE ONE NIGHT, after a bottle of Cab and a half dozen hits of weed, I decided to call Ruck, an old friend I hadn't seen in probably fifteen years. I still have a Polaroid someone snapped of him in the early 90s, when we were all at our most glorious. Surrounded by yellow light, he sits grinning on a motel room bed with a black 12-gauge Benelli across his lap: skinny, shirtless, barefoot, and tan. He could have been any of us back then.

In modern Facebook pictures, he still occasionally wore his patched-up Harley vest, but its buttonholes were at least sixty pounds from being reunited with its buttons. Ruck's trademark beard was still in place, but it was grey now, like my own. I shuddered clicking through the rest of his photos. Ruck in a burgundy recliner, beer in hand, three empties on the table beside him. Ruck seated before a fudge brownie with a single glowing candle—a birthday outing to the Macaroni Grill. Ruck in a lawn chair, a pudgy grandkid stumbling through a sprinkler in the background.

I found his number in my contacts, prayed it hadn't changed, and called. As the phone rang, I zoomed in on one of the images I'd found online, searching his ragged face for some trace of the devil I knew from years past, a man who I'd once watched shoot out someone's back tire for shorting him ten dollars' worth of crank.

"Jesus Christ, Ruck!" I said, as we took off running. "Bit much, don't you think?"

"Maybe so," he said, laughing. "But I bet he don't try that again."

By now, whether Ruck picked up or not, he'd see I'd called. Some late-night call from some old associate he hadn't heard from in ages. Well, to be fair, we were friends. Still, I've never liked asking favors, even from a guy who owes me one. You had to wonder if favors have expiration dates. Hell, even if they never expire, calling one in after too long makes you look desperate, like you have no connections of your own. And when he does pay you back, at which point you should be square, you somehow end up feeling like you're the one in his debt. It's the same every time. Thinking it over, I tried to remember why Ruck owed me a favor, and it occurred to me that maybe I was the one who owed him. Which only proved my point.

The call went to voicemail. I hung up.

"You fucking lazy bum," I said.

I looked at the clock on the oven. 2:17 a.m. I stood by the kitchen island, took a long, deep breath, tapped out a bump of coke on the granite countertop, and snorted. Fuck it. I called again.

This time, the call connected after the third ring. No one spoke, but I heard the faint sound of cowboys and Indians on the TV, a breath in and out.

"Ruck, it's Winston."

No reply except the sound of humid air.

"Ruck, wake up. I gotta ask you something. It's me, Winston."

"Hang on."

In two words, I recognized the gravel in Ruck's voice. The sound knocked me back a few decades in a blink. Fuck's sake. I don't care if it is a cliché. One day you're twenty, and the next you're fifty-seven. Funny how you can go twenty years without missing someone and then suddenly you miss him. Just like that. Ruck, my old friend. The time we'd spent apart suddenly seemed ridiculous, unthinkable.

Rustling sheets and a groan as he hoisted himself out of bed.

I knew from Facebook he'd remarried after Amy died—this time to a thin, short-haired woman who dressed like she'd bought her whole wardrobe at a Trump rally. I heard the opening of his bedroom door and the sound of him easing it shut. She must have been a heavy sleeper.

"Wince." Still whispering. "What's up, old buddy?"

"Hey man, who do we know who could teach someone a lesson for me?"

I heard his front door open and a screen door creak. Ruck lit a cigarette and took a draw. I was relieved to find he was still smoking, at least. "The kind of lesson you wake up from or the kind of lesson you don't?"

"If I wanted somebody slapped around, I'd do it myself."

"This guy's in Austin? I mean, you're still in Austin, right?"

"Yeah, I'm still here. He's here."

"Heard there's a bunch of protests down there all the time. Heard it's like the sixties. Bunch of fat lesbians fighting for late-term abortions and shit."

"Is that what you heard?"

"Yeah, saw something about it. I can't see living in a place like that."

"Well, I haven't been going to any protests, if that's what you're wondering. Haven't actually seen any either. Although I'm not exactly spending a lot of time at the state capitol."

"No, I don't guess you would. So, who's going to be learning this lesson?"

"Guy I used to work for. Not a connected guy, just some prick."

"Would he see it coming?"

"No, he doesn't know me like that."

"Sounds like he doesn't know you at all then."

"I worked with him for a while, so he thinks he does."

"Let me think a second."

While he thought, I took another sniff and lit a cigarette of my own. I stepped out into the darkness of my back deck, smoking and waiting for him to do what I'd already done at least a dozen times, reviewing a list of all the people we used to know, marking them off one by one. Teddy Ainsworth and Dave Fry died in motorcycle accidents. Arnold May in a car crash. Luke Weber by 9mm, Slim Hopper and Lefty Guzman by .22s. Darby Spencer fell off his roof repairing a leak. Guys we knew had died of lung cancer, sepsis, cirrhosis, suicide, and a housefire. The unlucky ones were in prison.

Ruck blew out a lungful of smoke. "I don't think I know of anyone anymore, crazy as it sounds. If it was anyone else asking, I'd tell them you're the guy to call."

"It could be someone from out of state. Money's not a problem."

"Yeah, that's right. You're a banker now, I heard."

"Wealth management. Seriously, who can we call about my problem?"

"It's the middle of the night, Wince. I got no ideas at the moment."

"Could you at least think about it and get back to me? I mean, between the two of us, we gotta know somebody."

"Sure, man. Call tomorrow if you want. We should catch up anyway. My wife works a little candy striper gig down at the hospital, nine to one. You can tell me about your problem. Maybe I'll think of a peaceful solution."

"Last thing I want is a peaceful solution."

I was about to hang up when Ruck said, "You doing alright? Other than whatever this is about?"

"I don't know, man. I got divorced again."

"Second divorces are even worse than firsts. Or so I've heard."

"Yeah, well it was my third, and it was definitely the worst of all."

"Well, damn. And that one was supposed to be the charm. You heard about Amy, right?"

"Yeah. I never pictured you outliving her, that's for sure."

"That's what she said right before she died."

"But you're happy now, right?"

"I guess so," Ruck said. "Pretty happy. You?"

I sighed. "Ehh," I said. "Been better. Been lots better."

"Huh," Ruck said.

It didn't seem like he had any more witticisms to add, so I said, "Hey, you should get back to bed. The new wife's going to wonder."

"Yeah, I should. Alright, man. Call tomorrow."

"Will do."

"And if you need to call late, call late. The wife sleeps like a rock."

I disconnected, took a final drag, and flicked my cigarette into the yard. Shit. Ruck had been a last resort.

I took out my dick and pissed over the edge of the deck. Pissing outdoors never gets old; it's one of the last best things about being a man.

In the kitchen, I did a goodnight snort off the counter, downed what was left of the wine in my glass, and took myself to bed. Pretty easy considering I hadn't changed from my bathrobe all day. It took about a quarter of a second to decide against brushing my teeth. And that shower could wait until tomorrow, too.

For a while, in an attempt to spur myself into action, I tried applying a point system to my life. I got twenty-five points if I got dressed in the morning instead of just slopping around in my robe all day. Ten points if I took a shower. Ten points if I brushed my teeth. Fifty points if I made my own breakfast instead of having fast food delivered. I knew I was setting the bar so low it was practically in the basement, but you did what you could. I always told myself I could redeem a zero-point day at the last minute. Maybe a miracle would happen, and I'd decide to take a nice, hot shower before bed. Hell, brushing my teeth wasn't that much to ask. At the bare fucking minimum, I could use the mouthwash I was supposed to be using for my gums. Even for that little shred of effort I'd be willing to award myself five points. Maybe ten. But I'd known from the moment I woke up that morning it was destined to be a zero-point day like so many others, which was why I'd retired the point system altogether.

Pissing in the yard had been a good move. That way I didn't have to go into the bathroom. I used to be a pretty vain guy, but I avoided the mirror lately. I sometimes looked at pictures of myself from fifteen years earlier. You wouldn't recognize me. I had a personal trainer back then. He was an idiot shithead and a criminal, but he was a halfway decent trainer. I used to get up at six every day and finish a three-mile run with a cigarette. I could remember stepping out of the shower and one wife or another taking hold of my bicep or grabbing my ass. I looked good, and it felt good walking around in that body.

Now, when I looked in the mirror, I didn't see the kind of bicep women reached for. And, like my father before me, I'd just about lost my ass altogether. Plus, over the course of a few short years, I'd somehow grown the belly it had taken my grandfather two decades to cultivate.

I always figured I'd get back to working out one day, but so far inspiration hadn't struck. My old trainer had gotten himself thrown in prison, the moron, but I could still find another one and get my ass out of bed in the morning. What I really wanted, though, was something a little more magical. If it came down to it, I wasn't above getting a little liposuction done again. Botox, chemical peels. While I was at it, I could get pec implants, cataract surgery, and one of those surgically installed dick pumps so you never have to worry about getting hard.

I lay in bed arranging pillows—one to cuddle on either side of me, two under my head, and one under my knees. I looked at the ceiling, strapped on my CPAP, and closed my eyes, feeling the rush of air in my throat.

When the sun was up, I could nap until the whole day slid out from beneath me. Sleeping at night, however, had become something of an issue. It wasn't the coke. A little before bed usually relaxed me. There was something about that long stretch of darkness that set my mind free in all the wrong ways. For years, I couldn't wait to go to bed. My head would hit the pillow, and I'd drift off into these amazing motorcycle dreams, riding endless wheelies through waving wheat fields, over snowcapped mountains, skimming across the landscape on my bike like a god.

These days, my dreams were filled with frustration. I dreamed about getting flats in cars I'd owned twenty years earlier. I dreamed of home invasions. Though I knew every gun in my house was loaded, each one I grabbed in those dreams only clicked in my hand when I pulled the trigger, the ammunition disappearing in a puff.

For as long as I can remember, I've put myself to sleep by fantasizing about either sex or torture. Lately, sex fantasies only dredged up memories of recent failures. So, naturally, my mind went the other way. It wasn't hard. I'd been a prodigy at revenge since I was a kid. Plus, I was a notorious grudge holder with a lifetime's worth of enemies. When I couldn't sleep, I imagined torturing them. Over the years, I'd developed the practice into a kind of meditation. I stripped guys naked in front of everyone they knew. Dragged them behind my car. Burned them with cigarettes and cut them with knives. I sliced off their cocks a millimeter at a time and cauterized the wound with a shovel I'd held over a flame. In my mind, I'd done worse than anything you've ever imagined to ex-associates, ex-wives, and ex-friends. For years, it had been my second ex-wife, the whore, who took the brunt of my imagined abuse.

But that night, like every night for the past year, when I lay back on my pillow, the face that materialized belonged to my former business partner, Bryce Winters. I looked at him for a moment, holding him there in space, my mind a Rolodex of possibilities. I thought of everything I could do to him, wondering where to start. I made it a point not to rely on the same old torture methods. Or at least to throw in some twisted new scenarios each night. I pumped myself up getting ready for what lay ahead, building up my anger the way you work up a hard-on. I could see his oily, round face, his overgroomed beard, his greased-back hair, and his tall, soft body in a blue, double-breasted pinstripe suit.

In a blink, Bryce was naked, tied to a wooden table in my garage, his arms and legs splayed, tears dripping down into his ears. I took a pair of pliers from my workbench and watched him react. The fear was all over him. A twitch in his jaw, his eyes darting to the restraints for the tenth time, looking for a way out that didn't exist. I listened to him whimper and beg a little.

"You really thought you could fuck me over?" I said.

"Wince, please! We're friends!"

"If we're friends, how come I'm still stuck in a fucking two-year noncompete?"

"Look, I'll sign the release. I'll sign whatever you want. Untie me, and we can talk."

"It's too late for that." I wagged the pliers before his eyes. "What happens next is your fault, asshole. Just remember that."

I never gagged Bryce. I liked hearing him cry and yelp, playing him like an instrument. In my fantasies, I had soundproofed the garage and lined the bottom of the door with rags, so not even his blood could escape. Sometimes I only had to work on him for a few minutes before I drifted off to sleep. Other nights, by the time I fell into a dream, nothing was left of Bryce Winters but a quivering pulp of meat and raw nerves.

# 2

I WOKE UP around noon the next day. I looked at the phone on the bedside table and contemplated it for half an hour before picking it up. Jesus Christ. Four missed calls: one from my daughter and three from my mother. Still in bed, I cleared my throat and called my daughter.

"Don't tell me you just woke up."

Sam, of course, had been up since five, running, studying, volunteering.

"Actually, I just got home from a meeting."

"With who?"

"Guy I used to work with. David. You don't know him. What can I help you with, honey? I've got a call in five minutes."

"You have a call at 12:37?"

I took a deep breath. Sometimes I had to remind myself that Sam was my favorite person in the world.

"The guy texted me and said he was going to call in five, that's all. Anyway, you know I have all the time in the world for you. What's up?"

"Did you talk to Grandma about the ring yet?"

"Listen, honey, I'm going to need a serious break on this ring business, okay?"

"Well, we can't get engaged without a ring, and you know she promised it to me."

"I'll talk to her. I've already talked to her once. You know how she is. It's a process. Focus on grad school, would you?"

"Okay, Daddy. I just want to be sure. You do want us to get married, don't you?"

After three failed marriages, I wasn't sure anyone should get married anymore. And I certainly wasn't too thrilled about my only daughter marrying a Venezuelan stoner with a part-time job at REI.

I told Sam I wanted whatever she wanted, and I'd figure out a way to get that ring out of Luann come hell or high water.

My mother called a fourth time while I was outside having my first cigarette of the day. I let it ring. If it had been something urgent, she would have texted. I went to the kitchen, did a bump, found the last clean pan in the house, and fried some eggs. Fifty points to start the day. I left the dirty dishes with the others in the sink, took a shit, and did another bump before returning her call.

I'm sure there are plenty of benefits to having a mother only sixteen years older than you are, but I could never think of them.

"Well, I was starting to think you were dead."

It was a polite greeting by Luann's standards.

I looked around at my house, surveying the takeout bags covering the countertops, the stacks of unopened mail on the breakfast table, the collection of wine bottles that wouldn't fit into the overflowing recycling bin, the little anthill of blow with the straw beside it. If my mother could have seen it, she would have had a fucking stroke.

At seventy-two, Luann possessed the kind of energy I've never seen in anyone who wasn't on speed. Talking to her, you half-expected sparks to spray from her ears. Retirement had done nothing to slow her down. In fact, now that making money was no longer her life's purpose, she was busier than ever with an endless array of projects and events.

Once, about twenty years earlier, I had worked for her, a period which led to several years of us not speaking. Now we talked every day, and when she finished telling me about the various household undertakings and social engagements that filled her day, the conversation always turned to me and all I was doing wrong with my life, how I had her worried sick about how fat I was becoming, how much wine I was drinking, how on earth I was going to get by when I still had almost a year left on my noncompete. Luckily—if such a thing can be called luck—our conversations of late increasingly turned to the subject of my father, a sweet and mellow Budweiser devotee whose post-retirement existence had left him rudderless and forgetful, his mental faculties winding down like a clock.

"How's Dad doing today?"

"Well, he messed up the downstairs television. I keep telling him to use just the one remote like you said, but he keeps insisting he has to use the other one to change the volume."

"You don't need both to change the volume," I said. "I showed you how it works last time I was there. I handed you the other remote and asked you to hide it. Did you not hide it?"

"I did, but he must have found it, because now neither of them works, and he's got me confused too."

"Oh, Jesus," I said. "God, give me strength."

It took forty-five minutes and about thirty dollars' worth of cocaine for me to walk her through fixing their entertainment system, after which she immediately launched into a thirty-minute rant about her lawn guys. How I kept listening is anyone's guess. The conversation played like a recording of dozens of others we'd had exactly like it.

"Yesterday I went outside, and they had a microwave oven plugged in on the porch. They were microwaving their lunches, Winston. Two of them were napping. I mean, it's totally insane. You never did that kind of thing when you were in construction, did you?"

"Me? No. We took a crockpot to jobs. Turn it on in the morning and by lunchtime, you've got stew."

"Stop kidding around. This is serious."

"Well, hire someone else."

"I've already fired every other lawn company in town."

That sounded about right.

"This is always the problem," she said. "You assume you've reached the bottom of the barrel when you fire an incompetent idiot. You always think it can't get any worse. Then it does."

I thought back on my marriages. "There's definitely truth in that."

I kept meaning to bring up my grandmother's engagement ring, looking for an opening, some chance to question her reluctance to follow through on the promise she'd made to Sam years earlier, but I couldn't bring myself to do it. I just wanted off the fucking phone. Finally, I invented an incoming call and pretended I had to hang up urgently.

"Call me later," she said.

I looked at my recent call list to confirm that we had, in fact, spoken for over an hour. That's when I saw the call with Ruck from the night before. Until that moment, I had forgotten talking to him. I didn't call him back, and he didn't call me either.

By six o'clock, I was sitting on the sofa watching *Lord of War* for the tenth time when my friend Jade called. I paused the movie but didn't answer. Jade's call meant Trigger Finger would be playing a show soon. Jade's boyfriend, Brian, was the lead singer of an 80s rock cover band. The guys in the band were my closest friends. Or they had been before Covid locked everything down. As far as I knew, the band and the old crew had gone back to hanging out as usual when the scare died down, but I stayed dark. Jade, angel that she was, still called to lure me out every time Trigger Finger played, which was at least a couple of times a month.

She called twice. Both times I let her go to voicemail. Both times she left messages.

"I'm leaving two messages because I want you to pay twice as much attention as you normally do. Tomorrow is Brian's fiftieth birthday. I know you're still on Facebook, so you probably know. Stay tuned for message two."

"This is message two of two. Trigger Finger is playing at The Smokestack tomorrow at nine for Brian's fiftieth. Your name is on the list. It's a big deal to Brian. I'd really like it if you'd show up. If you need a ride, I can come get you myself or have someone pick you up. If you want drugs, I'm sure we can find whatever you need. If you want to bring a girl, we'll all be nice to her. You're either becoming a hermit or deliberately avoiding us. Either way, I'd appreciate it if you'd buck the trend and come out. I'll call tomorrow to leave more messages. If you want to be super cool, you could indicate in some way that you've listened to this, and I'm not just talking into the wind. Even a thumbs up would be appreciated."

I looked at the phone a moment, groaned, and texted Jade a thumbs up.

I unpaused the movie. Stoned, with the TV on, I could almost switch off my brain. When I wasn't sleeping, which admittedly took up a massive percentage of my time, I had passed my days for more than a year by rewatching movies I'd seen hundreds of times. *Scarface, Training Day, Heat, Taken, Face/Off*. I liked movies about colorful bad guys who got away with it or died trying. I had memorized every episode of *Drug Lords*, although I was already well acquainted with everything they had to report. Since I was nineteen, coke dealers—Escobar, El Chapo, Griselda Blanco—had been my heroes, and I knew their stories the way some meathead would know the players on his favorite football team.

*Lord of War* was the kind of movie I liked, and since they didn't make enough of the kinds of movies I liked, I ended up rewatching the ones I did over and over again.

By ten o'clock, I was watching some Lifetime movie about a ditzy brunette who turns out to be pretty smart when she figures out her creepy husband is a serial killer. It was the next thing Netflix played for me, and I was too lazy to change it.

My phone pinged. Luis, finally. Thank God. I'd been rationing out what little coke I had left, waiting for him to make some mysterious pickup that had been delayed.

*Got your stuff*, his text read. *Come over.*

I pulled my T-shirt away from my neck and ducked my head to get a whiff. Not great, but hardly worth a shower. I grabbed the keys to the Tahoe and headed out. I paused in front of the house to light a cigarette and, like always, took a moment to look up and down my street.

The house I rented was in an upper-middle-class suburb with a bunch of six thousand square foot tract homes built in the 90s. The subdivision was designed to give the impression of a "community" with little parks and schools and stuff, but it was basically just a safe, marginally elevated suburb with shitty traffic if you had any dreams of getting downtown. Personally, I never would have picked it out, but my girlfriend after my second divorce liked the place.

Every house in the neighborhood came equipped with a three-car garage and a family of foreigners. Aside from a white pinko next door with a collection of libtard signs in his front yard, everyone around me was Chinese, Indian, Iranian, or some other race. My old neighborhood was all Jews. I had no idea I'd ever miss them so much. After six years in my current neighborhood, I still couldn't figure out how I'd landed in such a scramble of humanity.

My neighbors were always outside pushing their kids in strollers or speed walking, talking on their cellphones in foreign languages. Lots of elderly Asian types squawking to each other. They all checked their mail religiously at the neighborhood's mailbox station across the street from my house and liked reporting me to the HOA for leaving my trash cans at the curb.

Luis, my coke dealer, lived in a suburb to the east, twenty-five minutes away. It was pretty much the only place I went other than to the gas station to buy cigarettes. I could make the trip in my sleep. Luis's house looked like mine, shrunken by two-thirds. There were just as many brown-skinned people around, but they were all Mexicans in his neighborhood. His white pickup sat in the driveway. It was the nicest thing he owned, bought with the money left to him when his mom died a couple of years earlier. He got the insurance money and bought the truck the next day. Classic.

I knocked and walked in without waiting. Like always, the house reeked of weed, incense, and litter boxes. Luis was seated before the TV, bald on top with shoulder-length hair everywhere else. He wore sweatpants and a tie-dyed shirt with a pot leaf design in the middle. It was his usual style. At sixty-five, Luis was the most prolific stoner I'd ever met. He was like one of those teenagers who discovers weed and somehow forges an entire identity around it. Well, weed and cats.

"Cabrón!" he said.

"Hey, homes."

"Already got you all set up."

He'd laid out two fat rails and a red three-inch plastic straw on the coffee table before my usual seat, a brown leather sofa with built-in recliners.

"How's the new stuff?" I said.

The two lines vanished up my nose before he could reply.

A deep breath, and my ears began to ring. Good God. The quality made me realize how shitty Luis's old supplier was. And how much my standards had slipped to accept the garbage I'd gotten used to.

When I opened my eyes, Luis was watching me the way a chef peeks through the kitchen door to the dining room.

"Pretty fucking good, right?"

I eased back into the couch. "Hell, yeah."

Luis leaned forward and did a bump of his own.

"You gotta try this new weed I got, too. Crazy strong. I've been smoking it all day."

No matter how much he smoked, Luis always acted the same, apart from getting kinda chatty sometimes. I don't think I'd ever seen the guy sober.

Luis rolled perfect joints. Every time he handed me one, I paused to admire it. I took a small hit and blew it out.

"The last stuff gave me a headache," I said.

"Not this shit, man. This shit's like heroin. Makes you real calm. Takes your mind off things. Hit it again, man."

I hated to fuck with the coke high, but refusing would have been an insult to his whole Cheech and Chong culture. I took a hit and held it.

"How's your back?" I said.

"Eh, the same. Your knee?"

"It's alright. This cold weather ain't great for it."

Luis smiled. "Oh, I got you something."

"You don't need to get me anything."

He was always getting me things. A sweater I'd never think of wearing from Walmart when he noticed mine had a hole in it, some rolling papers he said were better than Zig-Zags. A month or so earlier, he had given me a DVD of a Def Leppard concert he'd come across at a thrift store. I thanked him profusely, never mentioning I hadn't had a DVD player in about a decade.

Luis was on his feet now, retrieving whatever it was from his bedroom. "It's that thing we saw."

"What thing?"

"You'll see."

He brought out a small cardboard box with the words "as seen on TV" printed on it. He thrust it into my hands. "Microwave it to make it hot," the box said. "Put it in the freezer to make it cold." It was some kind of beanbag.

"It's like for your neck or your knee or whatever. They had them on sale at Walgreens. You said it looked good when we saw a commercial for it, remember?"

I did not remember. "Hell yeah. Thanks, man. That's awesome."

Luis grinned. "No big deal, man. Just thought maybe you could use it when your knee is jacked up."

"This is great."

For years, I had been nothing more than a regular client of Luis's. I'd stop by, grab the shit, and go. But over the year of

Covid shutdowns, I got in the habit of settling in for longer and longer conversations. One of his cats took a liking to me, which Luis took as a sign I was a good guy. Luis was nine years older than me, but we'd both grown up near Corpus, back when Mexicans and whites hated each other just for being Mexicans and whites, so in a funny way we had something in common. It had always been a joke in my family to mispronounce the word "sandwich" like "sangwich" because that's how Mexicans talked in Corpus back then. Saying stuff like "get off the truck" instead of "get out of the truck." Luis still talked that way. "Turn off the joint," he'd say.

Luis was paranoid about getting Covid. For everyone else, he left the shit under the doormat. I was the only person he let inside during the shutdown, even though I was taking absolutely zero precautions and figured the whole virus was another overblown case of snowflake meltdown.

Luis was a small-time dealer, an enterprising retiree supplementing his Social Security. He only sold enough to pay for his own habit, and always to the same five to seven people. Sometimes, as customers came and went, the same faces with the same eccentricities, Luis's house felt like a sitcom, with Luis and me as the stars, and them as recurring guests. Drugs made everything feel like a TV show or a movie. Maybe that's why I liked being around them. The more drugs, the better. Luis gave me a place to be around them all the time, and, in exchange, I gave him my friendship and my stories, even if he'd already heard them all.

As we sat smoking the joint, the doorbell rang, and Luis heaved himself off the couch to check the cameras. The monitors were in his bedroom. It was ridiculous, really. He had a camera system like he was protecting the Mona Lisa, but half the time he forgot to lock the front door.

"Come in!" Luis called. Then, to me, "It's Shelly."

"Of course it is," I said.

"Aw, she's alright."

Shelly, an obnoxious blonde in her fifties, had been coming to Luis for nearly as long as I had.

She breezed into the room dressed to the nines, wearing a pencil skirt, a cream cashmere sweater, and bright red lipstick. Most of the tall women I'd known had spent their lives learning to make themselves smaller, but Shelly always wore high heels, a rarity in Austin.

"Oh, Winston! Of course you'd be here! I saw your Tahoe outside, and I thought, 'That's got to be Winston.' It's been forever."

"I saw you last Wednesday."

"You're kidding. I swear, this week has absolutely crawled."

"How's Danny?" Luis said. For some reason, he always made chummy inquiries about her husband.

"Oh, God, don't ask. Actually, wait, no, he's in a good mood this week. Some land deal or something he's been working on finally worked out."

Shelly was almost stately until the moment she opened her mouth, after which she gave the impression of being a grossly oversized seagull.

She turned and looked down at Luis, who was maybe five foot five with shoes on. "You didn't even give me a hug," she said.

"Oh," Luis said. "Sorry."

She bent down and wrapped her arms around him, pressing her cheek against his and making an "Mmmm" sound. I looked away. It always bugged me a little the way she carried on as if she and Luis were such dear friends, when really, she only came by to get coke, do a couple of his lines, and bounce. If she did bother asking Luis how he was doing, you could see her lose interest the moment he opened his mouth. Luis obviously didn't notice because he always seemed genuinely pleased when she dropped by.

Shelly released him and stood looking down at the top of his head.

"So, this new stuff," she said.

"Right," Luis said, "How much are you thinking?"

"Oh, I don't know. Maybe a gram."

"A gram? Girl, you'll be back here tomorrow. Besides, I'm telling you, you're going to want more than a gram of this new stuff."

"It is pretty fucking good," I said.

"Well, what should I get?"

"Wince is getting four grams."

"Five," I said.

"It's that good?"

"Best I've had since Miami in the eighties," I said.

"You're kidding."

She was already rooting around in her bag for the cash.

"I'm going out for a smoke," I said.

I usually went to a chair out back and let one or two of the cats roam around the yard while I smoked, but this time I stood on the front porch. A Mexican couple across the street was arguing in their yard. I knew enough Spanish to get the gist. Like always, I thought about how glad I was not to be married anymore.

When my cigarette was halfway smoked, Shelly came out of the front door and smiled at me. "God, I miss smoking," she said.

I held the cigarette out to her. "Want a drag?"

"I better not. Danny hates the smell."

"So, what did you end up getting?"

"Five. You only live once, right? It'll last months."

"How many you pay for?"

Shelly hesitated. "You know Luis likes to comp me a gram now and again. He doesn't mind."

"Huh."

She looked at me. "What?"

"Maybe from now on if you get five grams, you could pay for all five. I mean, since y'all are such good friends and all. Hell, if you're short and you need to borrow a Franklin or something, you can ask me."

"Well, aren't you sweet? A regular knight in shining armor."

When I went inside and sat down, Luis was in the kitchen opening cans of cat food.

"What did she end up getting?" I said.

"Five grams."

"Jesus. Quite the little junkie."

"I feel a little guilty though," Luis said. "I mixed it with some of the old stuff."

"Ha. You fucking crook. Hey, wait a minute. You told me you were out."

He shrugged. "I was pretty much out."

"Uh huh. I see how you are."

I sat on the couch and reached for what was left of the joint in the ashtray when my phone pinged. I looked at the digital clock over the TV. It was eleven thirty.

"Uh oh, Tonya's calling," Luis said.

I took my phone out of my pocket.

"Yeah, it's her all right."

Luis didn't exactly have the right idea about Tonya, but I didn't do much to correct his impression.

*Not far from your house,* the message read. *You home?*

*Will be in twenty*, I typed.

"Well, Luis, looks like I'm heading back to the crib."

"I knew it. You're too old to be so pussy whipped."

I stood, stumbled, and pulled out my wallet. "Gotta get it where you can. Gimme four grams," I said. "Of the new stuff."

Luis bagged them the way I liked, split up into eight half-grams. I dropped four hundred-dollar bills on the coffee table and grabbed the box with my beanbag in it.

"Shouldn't drive on that strong-ass weed. Do a line before you go."

"Not if I want my pecker to work," I said. "Don't worry about me."

# 3

BY THE TIME I pulled into my driveway, Tonya was already waiting out front, leaning against her old Camry. I bet the neighbors had their own theories about what a girl that pretty was doing outside my house at that hour. Their guess was as good as mine. I never understood why Tonya kept showing up like she did.

"You said twenty minutes," she said.

"I drive slow when I'm fucked up."

"Might as well just say you drive slow."

Inside, I poured us each a glass of wine while Tonya took in the mess.

"Seems worse than usual," she said. "You doing okay?"

"Yeah, I'm alright. The housekeepers come on Thursday."

"They skip last week?"

"Yeah, I switched them to every other week."

"Big mistake."

"Yeah, well, it ain't free."

I let Tonya take over the remote. She scrolled through the streaming channels looking for something we could agree on. I could never get over all the shit she hadn't seen. If she really was thirty-six, she hadn't even been born when half my favorite movies were made.

She put her hand on my thigh, and I took hold of it. Her nails were a little long that week, made up with French tips. Without taking her eyes off the screen, she brought my rough hand to her mouth and kissed the back of it.

When I'd met Tonya several years earlier, she was getting started at the escort agency I used from time to time. One night, celebrating my official divorce from my second ex-wife, the whore, I called and asked them to send someone over. Tonya—blonde, tan, and tight-bodied—showed up in a pair of jeans and a black silk blouse with no bra. Somehow, we'd never lost touch over the years, even through a third marriage and a handful of girlfriends.

I liked Tonya. She was good in bed, and she liked to kiss, which some of the girls wouldn't do because Julia Roberts wouldn't do it in *Pretty Woman*. She was one of those girls whose small tits only made her sexier. I saw other escorts too, but Tonya and I always ended up crossing paths. One night, catching my breath in bed, I told her about a deer I'd killed the previous weekend—a clean shot from sixty yards—and how much I was looking forward to getting the meat back from the processing place. I guess the way I described it was pretty good, because she said she wished she could taste it herself.

"I'm cooking it tomorrow," I said. "Come over. But if I'm the one cooking, I'm sure as hell not paying."

The next night, there she was, with a bottle of Cabernet, which she knew was my favorite. We ate venison, fried the way I like it. I kept her wine glass full and told stories to keep her entertained. She laughed her ass off when I told her about the time I won a live chicken in a jalapeño eating contest, and it made me laugh just hearing her. Tonya was good at listening, so much so that you found yourself talking more than you would with just about anyone else.

During Covid, things between us got more regular, even as the rest of my life became increasingly irregular. She was usually the one to initiate. I liked when she was there, and when I didn't feel like having her there anymore, I told her I wanted to go to bed, and she left. Of course, I didn't go to bed. I stayed up late, watched more TV, and did as much coke as I wanted free of judgement.

By that evening, when the two of us sat watching TV together, we hadn't fucked in probably three months. Occasionally, I went down on her for fun, but my dick hadn't been working so great lately. Last time she'd gotten me off was a blowjob that lasted an hour, the last forty-five minutes of which I'd spent repeating that I was about to cum. The bed was soaked in sweat, even though I'd just been lying there. On the way out the door, she made a show of rubbing her jaw like she'd taken a punch. That had been, what? Six weeks ago. Maybe eight.

"What did you do today?" she said.

"Talked to my mother for about two hours."

"I know you love that. How's your dad?"

"He's alright. He can't use a fucking remote control, apparently, but he's okay."

"What did you do after that?"

"Went and hung out with Luis."

"I thought you were slowing down on the coke thing."

"I have slowed down."

I tried to remember how many times I'd done a bump since Tonya arrived thirty minutes earlier. Four, I thought. Five, at most. Fuck it. I stood up and went to the breakfast bar and did another. I hate being told what to do.

I pinched my nose and looked at the ceiling. Lately, my nostrils hurt so much that I was using this steamer thing that was supposed to hydrate them, but it wasn't doing a very good job. "You've really never tried it?" I said. "I find that hard to believe."

"I think I already talk enough as it is, don't you?"

"That's true. You do talk a lot."

"Fuck you," she said. She got up from the couch, slipped into a pair of house shoes I'd abandoned nearby, and went in the direction of the bathroom.

"Where you going?"

"The bathroom," she said.

As soon as she disappeared around the corner, a racket rang out.

"Ow, fuck!" she yelled.

"What the hell was that?"

"This fucking gun with a knife on it fell over and cut my leg."

"Do you mean the bayonet? Come here and let me see."

"I'll show you when I get back from the bathroom. Call an ambulance."

Being chivalrous, I had a Band-Aid and an alcohol swab waiting when she got back. It was nothing but a scratch, but I knew it would set her off about the guns.

"What's a gun with a bayonet doing in the hallway anyway?"

"It's an SKS. It's a collectible. Old Russian gun, mothballed for decades and eventually sold to the Viet Cong to kill a shitload of Americans."

Tonya rolled her eyes, so I refrained from elaborating.

"What's it doing in the hallway?"

"Just minding its own business."

The truth was I couldn't remember how or why it had ended up propped against the wall in the hallway near the downstairs shitter. Probably I'd been tinkering with it—maybe a week ago, maybe a year ago—and got distracted and never picked it up again.

"I mean, one in the nightstand I can understand," Tonya said, "but nobody needs to be this well armed."

She spread her arms to encompass the entire house, where, it was true, firearms were strewn about in a fairly haphazard way. "I've never met anybody who likes guns as much as you do."

"That's because you only hang out with commies."

I've never been a gun collector in the traditional sense. I've never limited myself to weapons of a particular caliber, era, or manufacturer. Some guys only buy wheel guns or rifles or shotguns. When I had money, I bought guns. New guns, old guns. If it fired a bullet, I was at least somewhat interested in it. I liked having guns around. I liked how they looked, the way they felt in my hand. I liked taking them apart and putting them back together again. Most of all, I liked knowing what they could do for me if I needed them.

"Seriously, though, you're not doing anything with all these guns. Why don't you sell some of them like you said?"

"I never said I was going to sell *my* guns. I said I could build guns to sell."

"Whatever. Do that then. Seriously, what could we do to get you motivated? I mean it. You need a project."

"Honey, please, let's just watch a movie and chill, okay?"

"Come on, tell me what the first step would be."

"It's like I told you. Eventually, I'm going to set up a worktable in here."

"So, we need to find one online or what?"

"No, I have a table upstairs. It's just too big for me to bring down alone."

Tonya hopped up from the couch. "Well, let's go get it."

"I'll get my friend's kid to come by later this week."

"You always say you're going to do shit later, Winston."

"My back is kind of tweaked right now," I said. "I don't feel like moving that big fucking table."

"Isn't your generation supposed to be like the last of the real men or something?"

I sighed and closed my eyes. I didn't want to do it. I really didn't. But I guess I wanted to do it more than I wanted to argue about it.

I'd avoided the second floor for months, since my last girlfriend moved out and left the place in a state of such disarray it looked like the house of a regular loser. Junk was piled everywhere. Hard to believe I'd once entertained up there. I remembered a group of us playing guitars in front of the stack of Marshall amps, laughing and doing shots, each with our own bottle at our feet. The amps were still there, covered in dust, an artifact of a different era.

"No offense, but it's kinda horrible up here," Tonya said.

"Yeah, don't pay too much attention."

I didn't like seeing the place through Tonya's eyes. Hell, I didn't like seeing it through my own. A dried cockroach stuck in the carpet, a bunch of random tools, guns, metal ammunition boxes, guitars, a heap of bulletproof vests, a pile of T-shirts that didn't fit anymore, a wad of sheets from who knew when.

I could remember being proud of my house. For a long time, it had been pretty put-together, and people used to come in and say how great it was.

"Well, it's not the worst I've seen," Tonya said.

"Table's in the gun room."

I led her down a short hallway, opened the door, reached in, and flipped on the light.

Tonya squeezed past. "This is the famous gun room?"

It was really just a guest bedroom. One of four in the house. But I'd packed it with enough guns and ammunition to defend the Alamo, successfully this time.

I pointed to the corner of the room. "That's the table."

With its thick oak legs, it wasn't exactly my style, but many years earlier, my grandfather—a man I'd looked up to more than anyone who had ever walked the earth—had eaten most of his meals on its scratched, dinged-up surface. Now, it was heaped with random stuff. Headphones, sunglasses, a tangle of old phone chargers, a bunch of baseball caps and koozies. Most of it I'd forgotten ever buying.

My third wife got on a feng shui kick once. She used to say every room of your house is a manifestation of your inner life. It seemed like bullshit at the time, but I was struck in that moment, looking at the mess on all sides of me, how right she'd been.

"Fuck this," I said. "We'll do it another time."

Tonya ignored me, like I knew she would, and started clearing a path to the table. A half hour passed before we were able to wrestle it into the hallway. I slammed the door behind us. Together, we had achieved something I'd have never thought possible. We'd left the gun room in worse condition than we'd found it.

Tonya was stronger than she looked. Stronger than me, maybe, in my diminished state. Together, we carried the table downstairs, her walking backwards. By some miracle, we only dinged the stairway wall once.

I wasn't the only one breathing hard by the time we set the table down in front of the sofa. I fell into my usual spot, sweat pouring from my forehead. I glanced up at Tonya and saw her watching me mop my brow with a dishtowel I found stuffed between the cushions.

These sweating fits were something that had been happening more lately. I couldn't remember if it had started last year or the year before. The part that worried me was that the flood didn't stop when my pulse got back to normal. Even if I took a shower, the sweat kept coming. The only thing I could do was wait it out.

Tonya smiled and knocked on the tabletop like she was testing its sturdiness.

"If all else fails, you can play solitaire while you watch your drug shows."

"Great. Can we watch a movie now?"

Tonya sat down beside me like she always did and nuzzled up against me despite the growing puddle.

Looking back, I think it's always like this. Something small happens. You move a piece of furniture from one room to another. It seems insignificant then, but later you remember it as though you've rolled a boulder to the crest of a hill, where it teeters, loaded with inevitability.

# THE FIFTEEN-CENT BULLET

I remember when I was in the second grade I went deer hunting with my grandfather. I'd been hunting plenty of times before, but it was then, at seven years old, that we went out with the understanding that this would be the first time I killed a deer of my own. It was a rite of passage in the Fisher clan, and I grasped the gravity of it even at that age.

Before we left his house, my grandfather asked which of his guns I wanted to take. I knew he expected me to choose his old Browning, but I made a more surprising selection instead.

I pointed to the wall where he had hung his Canadian Centennial Winchester .30-30.[1]

"That one?" my grandfather said. "You sure?"

"Is that okay?" I said. I didn't mention I preferred shooting the 30-30 because I knew the kick would be so much lighter than the .30-06.

"I guess so. I've only shot it to sight it in."

"But it shoots good, right?"

"Course it does. It's heavy, though. And long. You think you'll be alright carrying it?"

"Sure, I can carry it."

"Alright then. Let's get her down."

An hour later, the two of us were sitting in silence in my grandfather's deer blind, which looked like a poorly built treehouse about ten feet in the air. Even though I was a hyper kid, something about hunting made me still. The gun in my lap, the feeling of being in the chilly outdoors. I liked sitting alone with my grandfather. No chit-chat. We just sat together, waiting.

---

[1] In 1966, Winchester marked their 100th anniversary by releasing a .30-30 lever action rifle, the Centennial '66. The U.S. Centennials were a gussied-up version of Winchester's Model 94, but they'd done a tasteful job of it. Twenty-four carat gold plated receiver and barrel band, heavy, 26-inch octagonal barrel. The gun weighed eight pounds and looked sharp as hell. The Centennial '66 caught on more than anyone expected. People like me, people into guns, bought them up like crazy, not to shoot but to hang on their walls. Even my dad bought one and hung it up at our house. My grandfather coveted my father's Centennial '66. They were sold out everywhere, so he couldn't get his hands on one of his own. Then, in '67, Winchester released the Canadian Centennial. My grandfather wasn't going to be caught short again, and even though he was one of the cheapest men ever to walk the earth and had never met a Canadian, let alone visited the country, he bought one and hung it on his wall. Same as the U.S. Centennial, the Canadian was a lever action 30-30, but with a black receiver engraved with maple leaves.

Eventually, I spotted an eight-point buck walking in our direction. I was proud to notice it before my grandfather did. I tapped his shoulder and pointed.

There was no question between us that the shot was mine. I raised the rifle and rested it on the branch in front of me. When I aimed, I put the bead just above the deer's heart, behind its front leg. Finally, the deer quartered slightly away from me and stood still. I pulled the trigger.

The noise was enormous. The wham of the butt against my shoulder.

The deer's knees buckled, but in the next instant it took off running. I looked at my grandfather.

"Pretty sure you got him," he said. "Let's go see."

Before we got to it, we heard it screaming, thrashing around in the brush. The deer lay halfway on a fallen branch. Only its head and front two legs moved, pawing frantically, scrambling in agony or terror.

"Wonder how it got paralyzed after it ran?" my grandfather said, scratching at the stubble of his beard.

I didn't say anything. I was afraid I'd cry if I opened my mouth. I couldn't do that in front of him. Not over a deer or anything else. The closer we got, the more frantic it became, trying to right itself, drag itself, falling and thrashing again and again on the broken branch, making almost human sounds.

I lifted my gun to fire again, but my grandfather stopped me. "Don't waste the bullet. Give me the knife."

Fuck.

I knew the word even at seven. Fuck, fuck, fuck. I checked my belt, but I knew it wasn't there.

"I forgot it."

"You dumbass. Where?"

"The truck."

"Fuck's sake, the truck's a half mile away."

My grandfather sighed. "Alright," he said. "Wait here. I'll get it."

"I'll get it," I said.

"Nah, you'll take too long. Just wait here."

He trudged away into the trees leaving me alone with the deer.

Though my grandfather had barely glanced at it, I couldn't stop staring. I'd flubbed the shot and had hit it in the back. There wasn't as much blood as you would think. It was just a red hole with a little dark trickle in its brown fur. It was a mystery how it ran those steps after it was hit. Maybe the running was what had worked the bullet into its spine.

It opened and closed its mouth, its tongue jutting out, its dark eyes wild. I stood with it for twenty interminable minutes, maybe longer, apologizing and crying. I had never felt so much sympathy for anything or anyone. I wished I could take it back. I imagined what it would have been like to merely watch the deer make its way through the woods without ever pointing it out to my grandfather.

When I heard his footsteps, I wiped my face on my shirt. Soon I would have to watch him cut its throat. I tried to brace myself for it.

# 4

THE DAY AFTER Tonya and I moved the table, I woke up a little after noon to two missed calls from my mother, a text from Sam, and a Facebook quiz Luis had sent at three a.m. titled "What's Your Power Animal?" I sat in my robe on the front porch, scrolling and having my first cigarette of the day. Heading back to bed for another hour or two of sleep, I noticed the bayoneted SKS lying on the hallway floor, where it had landed after scratching Tonya's leg. I started to walk past, figuring I'd deal with it later, but the thought of her finding it unmoved the next time she came over made me detour to pick it up.

Carrying the rifle like a soldier with its bayonet extended, I pictured Bryce Winters, holder of my noncompete, standing before me. I imagined charging towards him, ramming the blade into his stomach and feeling his weight transmitted through the gun's wooden stock, watching the expression on his face as I twisted and drove it upwards, shredding his intestines.

Realistically, there was no way I was going to carry the SKS all the way upstairs to the gun room, but I did at least lean it against a console in the living room where it would be less likely to fall. It was the kind of thing Tonya would notice and appreciate, whether she commented on it or not.

After a nap, another smoke, and a couple rails of coke, I ordered Whataburger and turned on the TV. Sometimes, sitting on the couch, I thought of the desk calendar I'd once used at Keystone Wealth. How many appointments, written in my assistant's perfect handwriting, would I have scratched through by two o'clock on a Wednesday? Seven? Maybe eight.

I touched the top of my grandfather's table, which now sat before me, a slightly—but only slightly—incongruous addition to the room. I'd brought it with me everywhere I'd lived in the nineteen years since he'd died. When I was a kid, I sometimes daydreamed about having special glasses that would let me see every place on earth no human being had ever walked. Now I wondered if there was a place on that table where my grandfather's fingerprints still lay undisturbed.

At that table, he had taught me how to clean my first gun, a bolt-action Marlin .22. By the time I was five, I could recite his stories of moonshining and backwoods fist fights he and his brothers used to get into. It sounded like a dream life full of outlaws who wouldn't call the cops no matter what kind of jam they got into. I never stopped being jealous of the era he was raised in, when you could get away with anything, when states, even counties, didn't communicate with each other. You could have a warrant in Williamson County, but as long as you never got pulled over there, no one was the wiser. Back then, you could drive to someone's house and throw a Molotov cocktail through the front window without it being caught on fifteen doorbell cameras.

I looked at the rifle in the corner, at its glittering bayonet, and remembered what Tonya said about my liking guns more than anyone she'd ever met. She hadn't meant it as a compliment, but I took it that way. Whatever you might say about me, one thing is absolutely true: very few people know more about guns than I do.

For instance, here's something you probably don't know. According to the Department of Justice, there are 175,977 "legal" machine guns in the United States, a number that only ever shrinks and never grows. That's not counting weapons owned by the military or the cops. I'm talking about machine guns available for purchase by civilians.

Machine gun means fully automatic. You pull the trigger once, and bullets keep coming out until you stop pulling. That's all it means. They're not magical killing machines or heat seeking missiles, and believe it or not, they've seldom been used in the commission of crimes. Still, because of an idiotic law passed under Reagan of all people, civilians can only buy, sell, or own fully automatic guns made prior to May 19, 1986.

In theory, any one of the 175,977 machine guns can be legally acquired by your average non-felon, non-domestic abuser, non-mentally defective citizen, but they're a pain in the ass to obtain, they require a tremendous amount of paperwork, and possession of one puts you on the radar of the ATF, who are the last people I want knowing the details of my home arsenal.

Before the Firearm Owners Protection Act passed in '86, the average guy on the street could walk into any gun store and buy an entry-level machine gun for around two hundred dollars. These days, you'd have a hard time finding one for twenty grand. Which means that almost every legal machine gun winds up in the hands of a well-off firearm enthusiast, who, likely as not, will take it to the range, mag dump a few hundred dollars' worth of ammo in fifteen minutes, post some pictures online, then go home and put it in their gun safe.

Still, people want automatic weapons. They have a cachet among criminals and regular gun-loving Americans who think they're fun to shoot. Which, of course, they are.

The moment you tell a guy who's into guns he can't have something is the same moment he starts doing everything he can to get his hands on that very thing. Like drugs, people are always going to come up with ways to get full-auto weapons. Laws don't prevent anything. And the fact is, any run-of-the-mill, civilian AR-15 can be modified to function as a machine gun, essentially turning it into the same full-auto M-16 the military uses.

As much as lefty politicians love to spout off about how any kid could turn an AR into a full-auto weapon, it's not that simple. Converting an AR to full-auto requires someone with tools, training, and a fairly comprehensive understanding of how firearms work. Within that group, an even smaller subset of people is willing to risk breaking the law, even among the world of so-called "gun nuts," a term that has always gotten under my skin, if you want to know the truth.

This had been the business idea I'd been mulling over for years and had mentioned to Tonya. It would take less than a grand to build a decent AR-15 from scratch and convert it to a full-auto M-16. After which, I could sell it for between five and ten grand, assuming I wasn't too scrupulous about who I was selling to. And why should I be?

Whoever they were, my buyers would have to have balls. Even possessing an auto-sear that drops into an AR and converts it to full auto is a crime. That one little part is, for legal purposes, a machine gun. And if you also own a receiver or a gun that could be modified with the auto-sear, the Feds call it "constructive intent," and you're looking at a ten-year felony sentence.

Somehow, I never worried much about that part. Who was going to pay attention to a white, divorced, out-of-work wealth manager growing fat and living in a house three sizes too big?

I'd been thinking about it for years. Thinking I'd do it when I had time. Tonya was right. If she could fuck slobs for rent money, why shouldn't I arm them?

By the time my phone pinged with the news of my Whataburger's arrival, I had summoned the will to force myself upstairs to the gun room for the second time in two days, where

I picked through the wreckage and found the parts I'd need to get started.[1]

As I ate, my phone pinged again.

Jade.

"Am I sending a car, or are you driving yourself?"

Trigger Finger. Brian's birthday. Fuck.

---

[1] Here's one more thing most people don't know about guns: the AR-15 is more than a weapon. It's a platform. Like Legos. The parts from any AR-15 are interchangeable with the parts from every other AR-15 in the world. It doesn't matter who made them or when. You can build one with shitty parts for hundreds of dollars, or you can spend thousands using the best parts the world has to offer. Eugene Stoner, who invented the AR-15, was the Eli Whitney of modern weaponry. He was working for ArmaLite when he designed the gun, pulling the best ideas from about ten other weapons and simplifying them. The "AR" in AR-15 stands for ArmaLite, not "assault rifle" or "automatic rifle," which a staggering number of idiots still seem to believe.

# 5

I SPENT MOST of the afternoon in bed, sleeping and dreading the show, wondering what I would wear. I had put on so much weight since the beginning of Covid, none of my clothes fit anymore. Finally, I got up and began trying on shirts until I found one that didn't look too bad. The ones I'd rejected formed a pile on the floor of the walk-in closet, and I left them there. I took off the shirt, did a line and got back into bed. I didn't know why I dreaded going out so much lately. For years, I'd loved going to Trigger Finger shows.

At a quarter to nine, I got into the shower. My first in four days. I put on the one decent looking shirt I'd found, the only pair of jeans that fit, and tracked down the motorcycle boots I wore to shows. I sat on my sofa and turned on a Liam Neeson movie Netflix had picked out for me. I could feel myself deciding not to go. I told myself that if Jade texted again, I would go, but only then. Two minutes later, the phone rang.

I sent Jade to voicemail and replied by text.

*Walking out the door. Got stuck on a call.*

The Smokestack, where the band played at least once a month, was five minutes from my house. I gave my name to the guy working the door, a tall, fat guy in suspenders and a fedora. There was a time I knew everyone who worked the door there, but I didn't recognize him, and he didn't recognize me. He searched for my name on a clipboard and let me in.

"Band's playing out back tonight," he said.

It was crowded inside. At the bar, I ordered a shot of tequila and a Michelob Ultra from a hot girl in her twenties and scanned the bar for familiar faces.

Out back, I spotted Brian immediately, wearing ripped jeans and a black Kiss T-shirt with the sleeves ripped off. His blonde hair was longer than the last time I'd seen him, down past his shoulders. He was holding a coil of cable, his own roadie, as always.

I called his name, and he turned, dropped the cable, came over without a word, and put his arms around me. He smelled like whiskey, and I was immediately glad I had shown up for him.

"Holy shit, man," he said. "It's great to see you."

"You think I'd miss your birthday?" I said. "I have a gift in mind for you, but I don't have it yet."

"You showing up is a gift, man. Don't get me anything. We're about to start. Jade's over there. She's been worried about you. Stick around, and we'll catch up after."

Jade sat at a high-top on the other side of the stage, which was really just a flat concrete patio with a shed roof. She was smoking one of her long cigarettes, holding court with the band members' girlfriends, looking like a combination of 70s Stevie Nicks and 80s Tawny Kitaen. When I interrupted her, she didn't make a big deal of my arrival. She hugged me and said, "Don't you look nice tonight? I saved you a seat." Real nonchalant, which was Jade's style.

I went around the table hugging the other girls and complimenting them on their cleavage, as, based on their attire, was obviously their wish. They smiled and laughed. When the band finished setting up, Brian thanked everyone in the crowd, and someone shouted, "Happy birthday, man!"

Brian ducked his head and said, "I hope I can still kick a little ass at fifty. I guess you guys are about to find out."

Then he turned around and motioned to the band, and they launched into "Livin' on a Prayer."

Jade and I shouted to each other over the music.

"Brian's been resting his voice for a week," she said.

"He sounds great."

Nothing had changed. We were seated close to Mark, who operated the lights. He bummed a cigarette from me and, when I gestured for it, he handed over the set list. I could recite the lyrics to every song by heart, even the ones I didn't like. I tapped out a bump on the table knowing no one would care, got Jade's attention, and asked if she wanted some. She shook her head. I didn't know if there was a hint of disapproval in her expression or if it was the usual maternal concern, but she never made a big deal out of that stuff.

I scanned the audience. It was the usual assortment of old people. People older than me. I'd heard for years that 80s music had experienced a resurgence among young people discovering it for the first time, but I rarely saw anyone under forty-five at

Trigger Finger shows, and honestly, most of the people in the crowd were pushing sixty or more.

In the crowd, I spotted Nikki, a friend of the group. I had flirted with her for years, though our timing had always been off. Either I had been involved, or she had been involved. Nothing had happened between us other than a brief parking lot blowjob one night after a show.

Nikki was in her mid-forties, but she still looked pretty good. She was the epitome of the cheap, trashy rocker chick I'd been drawn to since childhood. I'd always thought of her as sexy, though both my second and third wives found her annoying, and both had noted what they described as her "gummy smile." She wore a black bustier, and the tops of her breasts made oceanic waves as she toddled over to me on five-inch heels.

"Thank God you're here," she shouted into my ear, yelling over Brian singing "Girls, Girls, Girls."

"Thank God why?"

"I've been thinking of you all week."

I put my arm around her waist and smelled whatever knockoff perfume she'd dabbed on her tits.

"Oh yeah," I said. "How come?"

"I'll tell you in a minute." She moved my hand from her waist to her ass. "Don't leave without talking to me first, okay?"

She withdrew a pack of Capri cigarettes from her fake leather clutch. She took one out and put it in her mouth in a pretty sexy way, coating the filter with dark red lipstick. She turned to face me, and I lit it for her.

She inhaled, blew the smoke in my face and smiled. I looked at her gums. I could see what my exes were talking about.

It was rare to see Nikki without her ex-boyfriend, Dustin. Supposedly, he was her designated driver, but nobody had seen him sober in years. I looked up and spotted him out in the crowd. He was looking right at me. I lifted my beer in his direction.

"You and Dustin back together?"

"Fuck no. You know I just don't like to drive fucked up."

It was true she already had two DWIs, and a third would send her to prison.

"Want some coke?" I said.

"Are you kidding? Where can we go?"

"We don't have to go anywhere."

I tapped out a bump on the table and handed her a straw from my shirt pocket. I noticed the girls at the table watching her from the corners of their eyes. Nikki glanced at them, then lowered her head and inhaled the bump. She had fucked most, if not all, of the band members over the years and was generally unpopular with their girlfriends, despite her attempts to befriend them. She motioned for more coke, so I tapped out another little bump for her, then took the straw and did another myself.

"Oh, my God, baby," she said into my ear. "You always have such good shit." She melted against me, pressing her chest into my side. I put my arm around her waist again.

The attention from Nikki wasn't unexpected, and not only because we had flirted in the past. In the same way that women of her sort were my particular flavor, I was theirs. Something about my persona and my reputation with these girls meant I could get laid without a lot of effort when I wanted to.

During the break, I excused myself from Nikki, bought two shots of Jack Daniel's for Brian's birthday and took one to him. We did them together, and I told him how great he sounded. When he got pulled in another direction, I shook hands with the other band members and told them it was the best I'd seen them play.

On my way back to the table, Nikki's ex, Dustin, walked up to me and said in his gloomy way, "Hey Wince, long time, no see. How's it been going?"

"Not too bad. How about you?"

"Oh, not bad. Got fired the other day. Did Nikki talk to you yet?"

"A little."

"Yeah, we were hoping you'd come tonight. There was something she wanted to talk to you about."

That's when Nikki materialized and threw her arm over my shoulder. "Come with me," she said, completely ignoring Dustin. "Where'd you park?"

I figured she wanted more coke, and it was true it was more civilized to do it outside than off the high-tops at the Smokestack. I led her out to the Tahoe, and she leaned against me the entire way.

As soon as I sat in the driver's seat, she leaned over and began unbuttoning my pants.

"I don't think I'm going to be able to do anything," I said. "I've been partying a lot the past couple of days."

"We'll see about that."

It seemed like every woman I met thought she was incredible at sucking dick when most of them were mediocre at best.

She paused a moment after unzipping my pants, but she didn't slow down for long.[1] She put her mouth over the head of my dick and began sucking for everything she was worth.

As always, I tried fantasizing about things that would turn me on, aware all the time that on multiple occasions in the past I had gotten head from someone else while fantasizing about Nikki. I experienced a surge of optimism when, for a moment, my dick threatened to get hard, but it soon decided against it. I think she would have continued for hours if necessary, but finally I said, "Honey, it's really not happening tonight. How about a rain check?"

"Rain check sounds good," she said, wiping her mouth and chin with a Wendy's napkin from the cup holder.

---

[1] For many years, I had been well-known among friends for being naked a lot. On boats, at parties, around my own house. People used to call me Naked Wince back in the day. As a consequence, among friends in various groups, my dick had taken on a kind of legendary status. It was fairly long even when totally soft, and, frankly, it was a good-looking dick. It was fun having a big dick. It seemed right to me that I should be well hung when other losers were stuck walking around with average little dicks. Gradually though, over the past several years, something about my dick had changed. It seemed impossible, but I grew increasingly certain that it wasn't as long as it had once been. Plus, it was getting fatter. At first, I thought I was imagining it, but eventually, feeling ridiculous, I brought it up to my doctor who gave my cock a good long inspection and diagnosed me with something called Peyronie's Disease. It isn't really a disease. It's an accumulation of scar tissue that changes the shape of your cock. The doctor said I hadn't done anything to cause it, but thinking of all the abuse I'd put it through, I sort of doubted he was right. It was still a perfectly nice dick, and it still definitely would not qualify as a small when it was hard, but the weight gain and coke hadn't helped, so when Nikki pulled it out of my pants, or I should say when she opened my pants to find it, there might not have been as much to discover as she recalled.

The front of my underwear was soaked with saliva when I did up my pants.

"What did you want to talk to me about?"

"I have a problem I need your help with. I have a roommate who won't leave."

"A boyfriend?"

"Ex-boyfriend."

"Does he know he's your ex?"

"I've told him, but he doesn't want to hear it. Whenever I say something about how maybe we need to change things between us, he loses his shit, and now he's even started getting physical with me."

"Physical how?"

"Like grabbing my arm." She gripped her upper arm above a faded barbed wire tattoo. "I had a bunch of bruises last week, but I guess they're mostly gone now. He slapped me like three times once because he said I was too drunk, but I know that's bullshit, because I've been drunk around him like a million times, so why would he hit me that one time? And then one night he pushed me off the bed in the middle of the night for no reason at all. I think he's like a literal sadist. Like, I'm afraid he might actually end up killing me or something."

"So, what do you want me to do about it?"

"Well, I was thinking you could kick him out for me."

I rolled down the window and lit a cigarette. Nikki reached for it, and I let her take a drag. She tried handing it back to me covered in lipstick.

"Keep it," I said. I lit another. "He bigger than me?"

"Not by much, if he is."

"How old is he?"

"Forty-four. See, that's another thing. He lied about his age. He said he was thirty-eight."

"He work?"

"He's in a band. Liver Pill. They're kind of getting big in San Antonio."

"How long since you two fucked?"

"Last night. But before that it was like a week or something."

"We should go back in," I said. "I don't want to miss the last half of the show."

"But you're going to help me, right?"

The band was already playing when we got back. Jade pulled me close. "Hey, be careful, would you?" she said.

"I'm always careful, babe."

"Yeah, right. I didn't invite you so you'd get caught up in some drama with Nikki."

"There's no drama," I said, but when I looked around, I saw Nikki talking to Dustin in a conspiratorial way.

"Where are we going after?" I yelled to Jade. "Y'all hosting tonight?"

"No plans for later," she said. "Brian's an early riser now. He's got Pilates at nine. Then the gym. He goes every day."

"You're kidding."

"For almost a year. He hasn't had a drink in like two months. He's going to be so hungover tomorrow."

I looked at Brian gripping the microphone, his head back, his elbow pointed skyward. His bicep and deltoid were defined.

"Next thing you know, he'll quit smoking."

"He already has," she said. "Five months ago. He's insufferable."

I looked down at my shirt. Fucking thing made my stomach look huge. I wanted to tap out another bump on the tabletop, but now I wondered if everyone else was clean and thought I was some scumbag stuck in the past.

"Did fifty freak you out?" Jade said. "We went to Cabo, right? You didn't seem worried about it at the time."

"I can't even remember."

When they finished their set, I bought another couple of shots, and Brian did one with me. I worried maybe he drank it only to be polite, but he seemed grateful and happy. I told him how good he looked, and he beamed. Even covered in tattoos and muscled up, anyone could see he was an utterly pure spirit, one of the nicest, most honorable guys I'd ever met.

After a minute, he got pulled away again. Everyone wanted to talk to Brian on his big night.

"I think I'm going to head out," I told Jade.

"Oh, don't go yet. They're going to bring out a cake."

"Thanks, but it's been a long week. Tell Brian I'll call soon. We should all grab dinner or something."

"That would be great," she said.

Before I got to my house, I got a text from Nikki. *I thought you were going to talk to me before you left.*

*We can discuss it tomorrow,* I wrote back.

# 6

THE NEXT AFTERNOON, I awoke to texts from Nikki, Sam, and my mother. I already knew I was going to do whatever Nikki wanted me to. I didn't want to, and I didn't know why I'd end up doing it, but I knew I would. All night, I had slept badly, dreaming of holding a gun to the head of this long-haired man in a band with a stupid name and telling him to get the fuck out and never come back.

I texted Nikki and told her I would call later.

I texted my mother and told her I would call later.

I texted Sam to tell her I would deposit the requested money in her account, and that, no, I had not forgotten about ordering her new laptop. It had been a busy couple of days, and I would do it today.

I didn't want to get out of bed. I ordered some breakfast using my laptop and fell back to sleep. When I woke up two hours later, I knew the food I'd ordered was cold on the porch in a grease-stained bag. I ordered something else. When the alert pinged on my phone, I got up and put on my robe. It was three in the afternoon.

After I ate, smoked a cigarette, and did a couple of lines, I ordered the MacBook Sam wanted and transferred fifteen hundred dollars to her account.

Then I called my mother and let her chew my ear for an hour.

Since I was already suffering, I decided to call Nikki next. She answered on the third ring and told me in a whisper she'd call me back. She was at work, but she had a break in twenty minutes.

"I've got to get out and grab a pack of cigarettes anyway. Why don't I come to you?"

The old paranoia was kicking in. If I was going to talk about committing an act of violence, I figured it was best not to do it on the phone.

The dental office where Nikki worked was in a wasteland of depressing, low-end storefronts. I sat in the parking lot smoking until Nikki came out and got in the passenger seat. She was wearing light pink scrubs, and her top strained against her rack. Her hair was pulled into a ponytail, and she had removed most of her earrings, leaving a row of holes up her ear. She looked pale, and she'd recently attempted to pop a zit on her cheek. She smiled at me, and all I could see were her gums. It had been years since I had seen Nikki in the light of day. She looked rough and older. She lit a cigarette and immediately began talking in a rapid way that made me assume she'd recently taken an Adderall.

"His name is Mike," she said. "Do you want to know his last name, or is it better not to tell you? I don't know how this stuff works."

"Yeah, tell me. I'll look him up on Facebook. Would be good to know what he looks like."

"He's not on Facebook right now. He got banned."

"What for?"

"Oh, he posted something about how Nancy Pelosi should be raped in her house."

"What a fucking idiot."

"He's really smart, actually. I mean, when I first met him, he took this IQ test online and—"

"What is it you want me to do, Nikki?"

"I don't know. What do you think you should do?"

"I think I should stay the fuck out of it. How much furniture does he have at your house?"

"He doesn't own any furniture. It's all mine."

"Well, easy enough then. Pack up his shit while he's out."

"And do what with it?"

"Leave it on the porch and change the locks."

"No way. I need someone to be there. You don't know him. He's going to totally lose it."

"Lose it how?"

"Like *lose it*, lose it."

Like most women, Nikki could make you feel like you were going nuts just trying to have a conversation with her. At last, I managed to pry something useful out of her. The asshole, Mike, was out of town at the moment but would be back from his glorious tour of San Antonio later that night. He always got in late, she said. Usually drunk.

"We should do it tonight then," I said.

"You want to do it tonight?"

"I don't want to do it at all, but if he's out of town, you can pack his shit, give him a hundred bucks for a hotel room, and I can be there to send him on his way. Most guys won't do anything if another guy is there."

"You don't know Mike. He doesn't care who's there. And, anyway, I'm not going to give him a hundred bucks. I'm behind on my rent as it is."

"His parents live in town?"

"His mom does."

"Well then, he can go to her place. What time will he get home?"

"Like two a.m., probably. They're coming back after a show."

"Fine. When you get off work, get some boxes or garbage bags or whatever, and pack up his shit. I'll come to your place, and when he gets home, we'll tell him to find another place to stay."

"What if he gets violent? Will you bring a gun?"

"Yes, Nikki, I'll have a gun. Will he have a gun?"

"No. He had one, but he pawned it. Oh, and he's scared of knives, I think. One time we were watching a movie, and this guy got stabbed, and Mike was all freaked out by it."

"Would he call the cops?"

"No way, he has warrants."

"What for?"

"Unpaid child support, I think. His ex is a total bitch."

Watching Nikki sucking on her Capri cigarette, talking in her fast, jagged way, I saw in my peripheral vision a ghost—a younger version of myself chomping at the opportunity to save a girl I'd always wanted to fuck. I knew why ***he*** would have helped her. But what I still couldn't figure out was why ***I*** was going to. It was like wondering why your leg jumped when the doctor hit it with his little rubber hammer.

"So, if he gets home at two a.m., that gives you plenty of time to pack his shit. I can be there at one."

"What if he gets home early? Can you come at midnight?"

"Fine. I'll see you at midnight."

I stopped at KFC on my way home, ate greasy chicken in the Tahoe, did some coke off the kitchen counter when I got to the house, and slept for a few hours.

# 7

LYING IN BED, I wondered if leaving Keystone was the event that marked the beginning of my life's collapse. Maybe that job, even with all of Bryce's bullshit, had been the hard shell protecting me from the world outside. Maybe it wasn't because of my third wife splitting. Since leaving Keystone, I'd been in quicksand. I hadn't seen a dentist. I only saw my doctor when I needed to refill the prescriptions for my crazy meds. I hadn't been to a grocery store in over a year. I was running out of money. Nothing that used to be fun was fun anymore. As much as I didn't feel like fucking with Nikki's mess, at least it was a mission.

A little before midnight, I did a line and took off. The house Nikki rented was a junky two-bedroom on the south side of town with a lot of deferred maintenance visible from the street. On the porch, a rotting post leaned against a ten-gallon pot filled with dirt and a dead lemon tree.

Nikki answered the door in a Scorpions T-shirt and ripped jeans. Even when dressed as if preparing to clean the bathroom, her tits were as prominent as if a spotlight were trained on them.

"You got his stuff packed?" I said.

"I thought I'd wait until you got here."

"What? Why?"

"What if he comes in and I'm packing all his shit alone?"

"You gotta be fucking kidding me."

I spent the next hour picking through some loser's dirty clothes and guitar straps, listening to Nikki chatter about how awful he was and how, if only this would teach him a lesson, then things might work out between them. Then she'd reverse herself and say he was a worthless piece of shit, who would never amount to anything and he certainly wasn't going to drag her down with him, because one day she was going to figure out a way to get rich, and he was always going to be a nobody. Unless his band really did catch on, in which case...

Occasionally, we stopped to do lines of coke, which kept me focused and her bouncing off the walls.

As we packed, I noticed a pair of dog bowls in the kitchen.

"What are these?"

"Oh, pack those. They're for Mike's Rottweiler, Diesel."

"He has a Rottweiler? Where is the dog now?"

"With Mike. He takes him everywhere."

"So, when Mike shows up, he'll have a Rottweiler?"

"Does that make a difference?"

I don't know why it had never occurred to me before that Nikki was one of the dumbest chicks I'd met in my life.

Every sound coming from outside sent her racing to the front window to part the bent mini-blinds to see if Mike had arrived. I could feel the adrenaline stored up and begging to come out, wriggling free in little bursts.

Nikki kept texting someone on her phone. "Dustin says hi," she said.

"Dustin? Where the fuck is he, anyway? Tell him to get over here."

"He thought you'd want to handle it yourself."

"Well, he was fucking wrong."

"Okay, okay. Chill," she said.

After more texting, she announced that Dustin was on his way, but she somehow made it sound like they were both doing me a tremendous favor.

"Get with the fucking program. Get off your fucking phone and pack this shit. Let's get ready for this asshole."

When we finally finished packing, we did some blow. As I raised my head from her kitchen table, we heard a sound outside. Nikki said it was probably Dustin, but she went to the window, and found that it was, in fact, Mike and Diesel.

"He's here," Nikki said. "Oh, shit. This is him. He's here. Wince, he's here."

I looked outside and saw a man with long, dark hair fetching a guitar case from the back of a white van. In the yard roamed an unleashed Rottweiler which, even by the standards of the breed, could reasonably be described as enormous. The dog went over to a tree and pissed longer and harder than I ever have in my life. He looked back at his owner, who shut the van door, slapped it twice, and shouted goodbye to the driver.

I pulled my Glock from the back of my pants and grabbed one of the black garbage bags we'd crammed with Mike's possessions. I'd already decided to do it, so I didn't have to think. I told Nikki to open the door. She did, and I walked through it. Mike looked up at me—a complete stranger coming out of his house—and blinked, which was longer than it took for the dog to charge.

It was like I was watching the scene from a balcony ten feet above. The dog galloping full speed, opening its mouth, dripping saliva, Mike's eyes bulging as they landed on the gun at my side. From the balcony, I watched myself raise the Glock.

First, I aimed at the dog's chest. For a half second, I held my sights there. Then, when it was no more than twenty steps from me, I changed my angle and fired directly in front of it. The dog came to a halt almost instantly, startled by the burst of sound and by the patch of earth before it exploding and spraying him with dirt.

"Did you shoot my dog?" Mike said.

"No, but if you don't get it right fucking now, I'm going to."

But Mike didn't have time to do much before Diesel began to charge again. This time I thought I really would shoot him, but again, I aimed, dropped the gun an inch, and fired in front of it, into the boards of the porch. Again, it stopped as the wood opened up and sprayed splinters into its face.

Then Mike was holding Diesel's collar.

"Pull him back," I said. "Into the yard."

Mike did. Diesel reared up on his hind legs, snarling, trying to get at me.

"If you let that dog go, I'm going to kill it. Do you understand?"

"Yeah, I hear you, man. What's this about?"

"You're moving out."

"Who says?"

"Nikki doesn't want you here anymore. You're a fucking deadbeat, you don't pay for shit, and you're acting like an asshole, so she wants you gone."

"Whatever Nikki told you, she's full of shit."

"I don't give a fuck about any of that. The only thing I care about is that you understand you don't live here anymore."

"Yeah, okay, but that bag's not all my shit. I have a bunch of stuff here."

"We've got eight trash bags. Apparently, you managed to make it to fucking forty-five years old without ever owning a suitcase. This is the first bag. You're about to get the rest."

I kept my aim on the dog. People always say you shouldn't point a gun at someone unless you're really willing to kill him. I wasn't going to kill Mike, but no matter what happened, that dog wasn't going to get a taste of me.

"Nikki, bring out his shit," I said over my shoulder.

She started dragging the bags outside, making a mound of Mike's pathetic collection of worldly possessions.

"You fucking bitch," he said. "You stupid fucking bitch."

"Fuck you, Mike. I hope you like being homeless."

"Both of you, shut up," I said.

I backed up to the door and turned off the porch light.

Nikki kept bringing out the bags one at a time, slowly, like she was enjoying it.

I, for one, was not having fun. My heart was racing. I wanted it over with. I figured after two shots, someone had called the cops. All it took was a nosy neighbor peeking out her window to see a bearded man with a gun raised.

"Seriously, what the fuck am I supposed to do?" Mike said.

"Go cry to your mommy, you prick," Nikki said.

"Fuck you, you whore."

Mike's voice was getting loud.

"Shut up, you two," I said. "Where you go is not my problem. Just get the fuck out of here."

"Yeah, that's a great idea, mister. I'll walk ten miles with eight garbage bags."

"Call an Uber," I said.

"I don't have Uber."

"Ask him why," Nikki said.

"Shut up," I said.

"You have to have a credit card or a debit card or a fucking bank account to have Uber. You can't be a total fucking loser scamming money from people who trust you."

"You're a fucking cunt," Mike said. "Fuck you."

Diesel, still snarling, was so fixated on me, he barely seemed to notice when Nikki went over to him and stroked his head. Then she raised a wooden shower brush she'd pulled from one of the bags and brought it down on Mike's nose. Mike grabbed for his face, and I knew he was going to accidentally let the dog go and get it killed in the process. But I have to hand it to him. He didn't. He held on to the collar with one hand and reached for his gushing nose with the other. He yelled so loudly, I lowered the gun and stuck it in the back of my pants while his blood painted the lawn and the plastic bags. Nikki ran over and stood behind me.

"The fuck is wrong with you?"

"He called me a cunt."

That was the moment Dustin pulled up in an old, champagne-colored Nissan Maxima. I never would have imagined being glad to see that moron. He was the one who would drive Mike and Diesel and their eight bags of shit anywhere but there.

# ONE MAN ARMY

I remember when I was ten, my parents went to a friend's house for the evening. I was watching TV when a loud, urgent banging on the front door startled me to my feet. Back then, the only person who came knocking at our door unannounced was my grandfather, but he went to bed early, and he never knocked that way. I stood still until, after a minute, another round of knocking came, even louder and faster. What did I have to be afraid of? I went to the door and opened it. No one was there. It was dark out, the night silent and cool.

I closed the door and locked it and stood wondering what to do. Then the banging came again, this time on the kitchen door at the back of the house. I went to the door and looked out the window. No one. I checked the lock.

I went to my parents' room and took the .44 Magnum from the drawer in my father's nightstand. It was big and it weighed a ton, a Ruger Super Blackhawk.[1] I'd held it lots of times when my parents weren't there.

Another round of banging came. This time on a window, hard enough I was afraid they might be trying to break the son-of-a-bitch. It wasn't the window of the room I was in. I couldn't tell where it was coming from.

I picked up the phone beside my parents' bed and called the house where they were supposed to be. I was relieved when my father's friend answered and passed him the phone.

I whispered. "There's someone here, banging on the doors and windows."

"Get the .44," my dad said. His voice was light, distracted. "Hey, look out!" he said to someone else. I could hear laughter in the background.

"I'm holding it right now," I said.

"Good. You'll be okay. It's probably just some kids fucking around. We'll be home later. You said you could be the man of the house when we're away. So, be the man of the house."

---

[1] Lots of people think of the Ruger Super Blackhawk as just another old cowboy gun. It has that vibe. But the old ones, like my dad's, were different. The early Super Blackhawks were held together with three screws, which is why people call them 3-screws, and the design had a flaw: when the hammer isn't cocked, it rests directly against the firing pin. Drop it or mishandle it badly enough and the gun can go off. Ruger changed the design in the early seventies and offered to retrofit the old ones with a transfer bar safety. Some people actually had it done. But it's cooler to have one of the originals, pure and unadulterated. Some people load them with five rounds, leaving the hammer on an empty chamber. Like my dad, I always loaded six and let the chips fall where they may. It's not like I go around dropping guns anyway. Anyone who does deserves what he gets.

I went to the kitchen where I could sit on the floor with my back against the cabinets. The banging went on in different parts of the house for several minutes. I pointed the gun into the empty kitchen until my parents got home two hours later. They were drunk, and my father scolded me for not taking out the trash.

# 8

DRIVING HOME FROM the situation at Nikki's, I felt extremely unwell. I couldn't stop sweating. It was dripping off my scalp and into my eyes. My palms felt slick on the steering wheel. A balloon of anxiety swelled in my chest, pushing against my heart and lungs so I could barely breathe. Every once in a while, I noticed where I was but couldn't remember how I had gotten there from the last intersection where I'd had the same experience a few minutes earlier. I couldn't remember how I'd parted with Nikki. Had I told her what to say if the cops showed up? I didn't know. I thought about calling her, but I could barely drive, and now that I thought of it, I never wanted to talk to her again.

Maybe this time I really was having a heart attack. I lifted my hand and saw a cigarette there. For a moment, I was too afraid to take a drag. Then I did it anyway.

I went through the scene in my mind and couldn't think of anything I had done wrong. Well, I had taken my eye off Nikki. Never trust a woman. When was I going to learn?

I pulled at the collar of my shirt.

I couldn't figure out why I was so bothered by the events of the evening. Nothing had happened. I kept telling myself that nothing had happened. It had ended the way it was supposed to end. He left, and no one got hurt. Or no one got shot, anyway. If the asshole's nose was broken, that wasn't on me. If Nikki could be believed, maybe he really did have it coming.

A younger version of myself would have gone home, had a drink, and laughed about the whole episode, already practicing the story I'd tell about it. That's how I thought of everything then, as a story to tell. Whether I won a fight or lost it, it was another episode in The Winston Show. But now who was there to tell anyway? Not my mother or my daughter. Tonya would listen but would think I was an idiot. Luis might find it interesting, but what would the point be anyway? Some tough guy boasting to his geriatric buddy. I didn't feel like a tough guy.

Driving back to my depressing, sprawling house with the trash piled in the garage and my robe stained with wine, this didn't feel like a story to tell anyone. It felt like something I wanted to forget.

I made it to my house. It was safe, the same as it had been the day before and the day before that. But when I unlocked the door, I still wondered if someone might be standing in the dark or hiding in a corner on the second floor. I stood in the entry and listened. No one was there. It wasn't a fucking spy mission. I splashed water on my face, did a line off the kitchen counter, swallowed two Xanax, poured myself a glass of wine, and went out to the back deck to have a smoke.

Suddenly, I felt I did want to tell someone the story, after all—just none of the people I normally talked to. Besides, it was after three in the morning, and there was no one I could call that late without scaring the shit out of them.

Except maybe Ruck. It had been less than a week since I'd called him looking for someone to kill Bryce. And he had said we should catch up.

Fuck it. I called him. This time, I went to voicemail twice before he picked up on the third attempt.

He didn't say anything other than "Hang on" until he was on the other side of his creaky screen door. He lit a cigarette.

"You call later every time."

"I can let you get back to bed if you want. It's nothing important. Just felt like talking."

"That's alright. What you got to talk about?"

"It was kind of a wild night for me. Old days shit."

"Oh, yeah?"

"Yeah, I had to shoot at a fucking charging Rottweiler for one thing. It was kind of badass, actually. Fucking thing was running at me. I fucking drew down on it like Jesse James or some shit."

"Niece of mine had a Rottweiler. Pretty cool dog. They're loyal as hell."

"I didn't shoot it. I just shot near enough to keep it from biting me."

"Oh. Well, that's good."

"Yeah, I mean, it felt like old times, you know? Doing all this stupid shit for a chick. Getting into some trouble again. I gotta get out in the world more, you know? I'm fucking turning into a softie. Like, I can't spend my life sitting around doing nothing all day, you know?"

Ruck yawned.

"You can go to bed, man," I said. "This ain't anything important."

"I got time, Wince. Tell me what happened. But slow down. You're talking a million miles an hour."

# 9

BY THE TIME I ran out of coke, I'd put my grandfather's old table to good use, building three AR-15s with parts I had on hand in the gun room. If I'd dug further and gone through some boxes, I probably could have put together two or three more. That's how much shit the gun room contained. It was like an old version of me had really believed this day would come.

In order to convert them to full auto, I still needed to mill out the inside of the receivers to install the auto-sears, which I also had on hand. For that, I'd have to track down a bunch of shit and spend a while in the garage doing precision work, and there was no way I was doing it without coke.

I'd called and texted Luis a few times over the past couple of days, but I hadn't heard back. It wasn't like him, and it bugged me. It bugged me more the lower my supply got. I checked the timestamp on his last message. It was that four a.m. "What's Your Power Animal" quiz from days earlier. Usually, he sent me that kind of shit all the time, cat memes and dumb jokes. I called him one more time and didn't get an answer. Surely, if he was in jail someone would have told me. I decided to head over to his place and see what the hell was going on.

When I got there, his truck was in the drive, exactly the way it had been the last time I saw him. In fact, I had the impression it hadn't moved an inch. I rested my hand on the hood. It was cold. I knocked on the door and tested the handle. Locked. I looked under the doormat and found a collection of hundred-dollar bills from people who had picked up whatever he had left for them. Normally he would have at least gone out to collect the money at the end of the night. It was spooky seeing the money there. I picked up the cash, folded it, and put it in my pocket. I turned and looked at the street. It was empty. I called Luis's number. Straight to voicemail again.

I went back to the Tahoe, opened the console, and found a Glock 19 I kept there. I put the Glock in the back of my pants. In the backseat, I found a long, flathead screwdriver.

I went around to the side of the house. I knew the gate would be locked without even checking. No dealer is going to leave his back gate unlocked. I checked the gate. It was unlocked. Fucking Luis.

The blinds at the back of the house were drawn. I took out my Glock and banged the sliding glass door with the butt. The cats rushed the window, parting the vertical slats and meowing up at me, but Luis did not appear. I took the flathead screwdriver and pried the door off the track.

The second I walked in the door, I noticed the house smelled wrong. Weed smoke and incense were gone. The only thing in the air emanated from three stinking litter boxes. The cats swarmed me, curling around my legs, even the one who didn't like me. Holding the gun, I filled their empty water dish in the kitchen sink. They ran to it nudging each other out of the way to get a drink. I stood listening to the house. The only sound was the cats lapping.

"Luis!" I called out. "It's Winston."

No answer.

I dropped the cash from under the mat onto the coffee table and went through the living room into the hallway. I walked past the open door to the guest bedroom. Nothing looked off. Same for the bathroom.

The door to the master was closed. I knocked and called out again. "Luis, you in there?"

This time I heard a sound, a sudden, human stirring.

Trigger discipline is a term you hear a lot if you're into guns. It means you keep your finger off the trigger, leaving it extended against the trigger guard until the moment you know you're actually going to fire.[1] Despite the sound and the generally eerie atmosphere, I kept my finger off the trigger and congratulated myself for doing so.

"Luis?"

The sound on the other side of the door grew louder and more urgent, like I'd awakened something in the bedroom. A wild, nonsensical groaning.

---

[1] When I was a kid, people didn't talk about much trigger discipline, because back in the day most people were carrying revolvers. You're talking about twelve to fifteen pounds of pressure to pull the trigger of your average double action .357. That's a lot of pull, a lot of intention. Then everyone started carrying Glocks, and trigger discipline became a much bigger deal. It takes six, maybe seven pounds to pull the trigger on a Glock. That six pounds makes a shitload of difference when some Black kid jumps out of an alley on his way home from school in front of a rookie cop.

"Luis, it's Winston," I said to the door. "If you're in there fucking someone, you better say so, because I'm about to come in blasting."

The only reply was more groaning, louder this time.

I opened the door and went in Glock first.

Luis definitely was not getting laid. In fact, seeing him naked and writhing on the floor, his little brown, uncircumcised dick waggling around beneath his fat belly, it was incredible to me that he had ever gotten laid in his life. He was on his back, halfway between the door to the bathroom and the bedroom, the carpet around him stained with piss. He jerked and bent and carried on making those crazy sounds. I'd seen ODs before but nothing like this. I tossed my gun on the bed and knelt down beside him.

"Hey! Luis, look at me."

His eyes fluttered open and closed, his pupils never lighting on anything for longer than a second.

There was a desk chair on wheels in his room. I pulled it over and lifted him onto it. He yelped and cried out, almost appearing lucid, then he zoned out again. There was a towel on the floor. I tossed it over his lap. He could barely stay upright. He kept moving, wagging his head around, babbling nonsense syllables.

I pulled open his dresser drawers searching for clothes I could get him into easily. I managed, despite his resistance, to help him into a shirt. When I tried getting him into a pair of sweatpants, I saw one of his legs was red and swollen. I touched it, and his entire body reacted, jerking away from me. But I couldn't leave him naked, so I yanked the pants up as quickly as I could, ignoring his squeals.

Time always slowed at moments like this. I noticed the details—the time on the bedside clock, the still padlocked door to the closet where he kept his coke in gem bags along with about fifteen strains of weed in labeled mason jars, his cell phone and wallet on the dresser. It was like being reminded of a hidden limb, the way my mind could go quiet.

I stood in front of Luis, holding him upright in the chair whenever it looked like he might tip over.

Finally, after debating the matter a moment, I took out my phone and called his middle son, Manny. Manny was a fuckup but, of Luis's sons, he was the one who lived closest.

"Manny. Winston here. I'm at your dad's house. There's something wrong with him."

"What's he doing?"

"He's fucking thrashing around going apeshit. He's incoherent. He was on the floor. He pissed himself. I think his leg is broken. I don't know how long he's been like this. When's the last time you talked to him?"

Manny paused. If he felt any urgency, he didn't show it. "I don't know. Last week sometime. What do you want me to do?"

"What do I want you to do? What the fuck do you want *me* to do? You want me calling the cops over here, Manny?"

"Fuck no, don't call the cops."

"Exactly. So, what the fuck? You need to get here."

"I could be there in probably like two hours."

"He could be dead in two hours. I don't know the last time he had insulin. He's fucking out of it, man. He's fucked up."

"Maybe give him some insulin. See if it fixes him."

Fucking Manny. He'd never been a terrible guy until he picked up an Oxy habit to go with his coke habit, after which he'd turned into an absolute piece of shit.

I hung up and called Chucho, the youngest of Luis's fuckup sons. He didn't answer.

I got Luis situated in the chair as stable as I could and started going through the house collecting everything incriminating. It was no small task. Within a minute of snatching up bongs, pipes, rolling papers, scales, empty baggies, and baggies full of coke, I had to go back to Luis's bedroom closet—I knew where the key was hidden—to find an overnight bag to throw it all in. I grabbed two grams, stuffed them in my pack of cigarettes, then took out two hundred-dollar bills and dropped them on the coffee table with the rest of the cash. I checked on Luis, who was still mostly upright, so I took the suitcase outside and stashed it behind an old, rotting shed behind the house.

Inside, Luis had managed not to fall out of the chair. I grabbed his cellphone and wallet from the dresser and started wheeling him through the house, the cats following along, meowing up at him as he babbled back at them.

When we were almost to the front door, I called 9-1-1. On the phone with the dispatcher, I opened cans of cat food and dropped them on the floor. The cats were on top of them before they hit the ground. I grabbed a bottle of water from the fridge, suddenly realizing the nonsense syllables Luis had been babbling could have been *agua*. I turned out to be right. When I held it to his lips, he sucked on it like a baby bottle.

I waited by the door with Luis until I heard sirens, then wheeled him out. By the time the ambulance pulled up, we were waiting at the curb, him still flopping around in his office chair.

When I told the EMTs how I'd found him, one of the guys said, "Dude, you saved your friend's life."

"I think his left leg is broken," I said.

They cut open his sweatpants. "Yeah, it's broken," one of the guys said. "Jesus Christ."

My last trip in an ambulance had been almost ten years earlier. Speeding down the highway on a Honda CBR 954RR, I let a couple of idiots on crotch rockets goad me into popping a wheelie. It was kind of my trademark at the time. I was the old guy still popping wheelies. A decade earlier, and I was already the old guy. I pulled off the wheelie, of course, but I hadn't counted on road construction. I hit a patch of gravel, lost control, and next thing I knew, I was lying on the shoulder of the highway, L3 and L4 shattered. A second later, my rearview mirror slid over and came to a rest before my eyes, bringing with it a picture of my daughter. I'd placed a sticker made from one of her class photos on the mirror to remind me never to do anything stupid.

When we got to the hospital, the paramedics wheeled Luis into Emergency, and I stayed in the waiting area. I texted Manny and sat in a plastic chair across from a guy in a high-vis vest pressing a wad of napkins to his bleeding forearm. The place was packed with the usual misery: a woman rocking and holding her stomach, a guy on the floor getting his brow mopped by his wife. The new arrivals, if they were smart, skipped the line by claiming they were experiencing chest pains, even if they obviously only had a sprained ankle. They always got bumped to the head of the line.

I watched the show for twenty or thirty minutes and scrolled Facebook for another half hour. Finally, Manny slopped through the automatic glass doors in a pair of grease-stained jeans and a T-shirt advertising Newport cigarettes. He looked as bad as I'd ever seen him. Like everybody who does uppers and downers at the same time, he was always fucked up. Too sped up or too slowed down. Even from across the room, it was pretty easy to see where he'd landed that day.

Finally, he spotted me. His eyebrows rose, barely taking his lids with them.

When he plopped into the chair next to mine, I saw that his nostrils were red, a trace of white powder ringing them. Lucky bastard. He'd been doing lines in the comfort of the parking garage, while I'd only managed to sneak a quick snort in a bathroom stall.

Manny turned his head and looked at me. "So, like, what's the deal?"

# DAD'S DEAD FRIEND

I remember when I was twelve, a bunch of us were walking through a field of brown knee-high grass, holding shotguns: me, my dad, two of his buddies, and two of mine. Dove hunting. I wore a blue corduroy jacket with a shearling collar. It was Saturday, and my friends and I were talking about girls at our school.

One of my father's friends started complaining about the little .38 he wore on his hip. He unholstered it and examined it. I could hear its metallic clicks. Next, I heard a shot and looked at the sky, but there was nothing to shoot at. I turned to see why someone had fired, and my father's friend was standing there with a little hole between his eye and his nose. The gun was still in his hand.

"I shot myself," he said. "Did I shoot myself?"

My father gripped his friend by the shoulders and looked at him.

"Yeah," he said. "You shot yourself all right."

"That fucking gun. How in the hell did I shoot myself? You think I'm going to die?"

He fell to his knees before my father answered, but he still looked alert and mostly normal. He said my father's name, which was also mine.

"I think you might," my father said.

I'd been out hunting with the man probably a dozen times. He had a daughter two years ahead of me in school.

He kept talking. "Wonder if we should head back and go to a hospital, or if that'll do me in faster," he said.

My father and his friends debated the question, but we were far from anything, and the man's speech was slowing down. Then he stopped talking and wet his pants. I remember I felt bad for seeing it.

He fell forward on his face, and my father felt for a pulse. My dad and his friends carried him back to the trucks, and us boys walked behind them watching as they wrestled his dead weight across the field. By then, he'd shit his pants, and they struggled finding a way to carry him. One of the trucks was his. My father fished the keys from his friend's pocket and drove us back to town. I sat up front with my father, and his dead friend lay in the bed of the truck.

# 10

THE NURSES SAW me at the hospital so often that first week, one of them asked if Luis was my brother.

"I look Mexican to you?"

"Well, I figured either that or you're boyfriends," the nurse said. "And you look more Mexican than gay."

It was the nicest thing anyone had said to me in months.

"We're friends."

Luis, in a moment of lucidity, perked up enough to say, "Don't listen to him. This guy is a homo for sure."

It turned out he'd been lying on his bedroom floor for about thirty-six hours before I found him. He'd tripped getting out of the tub, broke his leg, and couldn't reach his phone or his insulin.

It would have been nice if patching him up was as simple as giving him an injection of sugar water, slapping a cast on his leg, and sending him home, but once they got him stabilized, they found that a long-ignored kidney condition had arrived at a critical stage. Things had degraded badly enough during his time on the floor that he'd have to be on dialysis for the rest of his life unless they found a donor. I tried sounding optimistic when he told me, but who was going to give a kidney to an old stoner cokehead like Luis? In a perfect world, his sons maybe. Except theirs were probably as bad as his. Maybe worse.

Eventually, one of the nurses—a guy who definitely *was* a fruit—asked if I'd be willing to get tested to see if I was a match. Whatever drugs Luis had in his bloodstream, I had at least twice as much shit in mine, and I didn't like the idea of anyone poking around thoroughly enough to find out. Still, I went through with it and was relieved when I was ruled out as a candidate.

The hospital was where I started spending time with Luis's youngest son, Hector. He was the only one of the three who actually had his shit together enough to visit once a day. Unlike Manny and Chucho, Luis hadn't had much of a hand in raising Hector. After Luis knocked her up, Hector's mother left Austin and returned to the Rio Grande Valley to raise him with the help of her parents. Hector was eighteen when he tracked down Luis. They kept in touch, but Hector had stayed in the Valley until about five years earlier, when he moved to Austin.

He did share physical similarities with Manny and Chucho, but Hector was purebred to their mutt. Shitkickers love to iron their jeans, and Hector wore his with starched creases. He always wore pressed shirts, cowboy boots, and a trim, black mustache. He was respectable. He talked like a normal guy, no gangster bullshit. And he had the kinds of businesses criminals were supposed to have. He owned a laundromat and had a stake in a Mexican bar in south Texas. Based on what Luis told me over the years, Hector moved a lot of coke and had been doing it even before the two of them reunited. A gene passed down from his father, I guess.

I'd met Hector before at Luis's house, but I'd never sat in the same room with him for long. He always swept in with something to do and somewhere else to be. He'd drop off a load of coke or pick up an envelope of cash from his dad and head out. The way his eyes moved, you could tell he was always figuring something out, always working on the next thing, moving along to another appointment or phone call. Not exactly shifty, but definitely not settled.

At the hospital, I saw a slowed down version of Hector, a man who did as a son should. He canceled everything and sat. Luis was asleep for most of the first couple of days, so Hector and I had our first real chance to talk. Nothing especially important, just getting-to-know-you bullshit. And when Luis woke up, or was napping, or was watching TV, we continued talking, sometimes to Luis's annoyance. It was as if Hector and I understood each other but kept talking to verify we were correct in thinking so.

"What about the shit my dad got last time?" he said one day. "You tell any difference?"

"Definitely," I said. "Good shit. At least it doesn't smell like gasoline like the old stuff."

Not wanting to place blame, I didn't mention it, but I had begun to wonder if the quality of the product had been my trouble all along. Now that I was getting good coke, I had actually finished building three ARs. Nothing I would normally brag about, but compared to my dismal output of the last year or so, it was at least noteworthy. Who knows, maybe with good coke the rest of my life would turn around too.

"Yeah, that's what I thought," Hector said. "It smells right. I switched guys. That's why the new stuff. Gotta deal with more bullshit, but that's how it is if you want to get bigger."

"That's been my experience."

"My dad told me you've had your share."

"Of coke? Yeah, I guess."

"Experience. Dad says you used to be in the business."

"Oh. Yeah, I was."

"Why'd you quit? Get caught?"

"I quit because I had a kid. My ex was a piece of shit, so I couldn't wind up in jail and expect her to raise my daughter. So, I changed my line of work."

"Now you're pretty straight, right? My dad said you do something with money. Banker or something."

"Wealth management. I was doing that, but after Covid, I sort of had to take a break. I took a year off sort of fucking around. Now I'm waiting out my noncompete."

"How much time you got left?"

"Ten months. It was for two years."

"Shit, man. Two years they say you can't work? I'd figure out where they're keeping the contract and burn the place to the ground."

I liked Hector more every minute.

"Yeah, industry standard is twelve months. Fuckers had me over a barrel, and I was dumb enough to sign."

"So, what do you do then? Guy like you can't sit around and watch *Judge Judy* all day."

"You'd be surprised."

A nurse pushed through the doorway to look in on Luis. She checked his IV bags and monitors, then initialed his chart and left us alone.

"You said I'd be surprised," Hector said. "Not much surprises me. So, what do you do all day?"

What I meant was that he might have been surprised to learn that a guy like me actually *could* sit around and watch *Judge Judy* all day. But instead, I said, "Guns. AR-15s. Fully automatic. And silencers."

"You're fucking with me."

"No."

"Where the fuck are you getting full-auto ARs? Like from Russia or some shit?"

"I build them."

"Huh." Hector let that sink in. "That's pretty interesting."

"It has its moments."

"Listen, something I want to ask you. An associate of mine usually goes with me to make pickups, but he's out of town for a while. Actually, he's locked up. Dad went with me last time, but obviously he's out of commission, and, honestly, he's not that great at the job. I've got to make a run later this week. I'm still waiting to find out the exact day. It's nothing complicated. Just need someone to drive. I go in and get the shit, you sit there and keep an eye out. I've been to this place twice already, and it's gone smooth both times. But it's a no bullshit kind of place."

"Why would I want to do something like that?"

"Thought it might be fun for you. Like the good old days. Plus, I could kick you a half ounce for helping."

"You wouldn't rather have Manny or Chucho go?"

"If I gotta choose between picking up a bunch of coke with a half-retard cholo behind the wheel or with someone who looks like you, it's not a hard choice."

"Where would we be going?"

"Junkyard about an hour outside of town. Close to San Antonio."

I almost laughed. Fucking junkyards. Of course.

"Why wasn't Luis good at it?"

"Well, he's not very steady for one thing."

"How steady does someone have to be?"

"Steady enough not to smoke a joint and go into a coughing fit in front of the guards while you're supposed to be watching my six. That would be a good start. It's an easy job. We pick it up, take it to my dad's house, then you're done."

Hector looked at me and smiled. "Then maybe we can talk guns. If you know what you're doing, I'm pretty sure I can help you sell as many as you can make."

## 11

I'D BEEN DODGING Tonya for days. She texted, but I hadn't been in the mood for more company than I was getting at the hospital, and I told her so. I explained over text about the Luis situation, but I sensed she might be annoyed about my unavailability. It was sort of par for the course. Tonya and I had exchanged words more than once about my need for time alone, sometimes for a week or two at a stretch. Just because we'd hung out four times last week didn't mean I wanted to hang out four times this week. In fact, it probably meant the opposite. She always figured it had to do with the coke, which it didn't. Sometimes I just liked to be alone.

As time went on, she sometimes took my absences in stride, and sometimes I had apologies to make if I wanted things to go back to normal. We weren't boyfriend and girlfriend. I'd long since fallen out of the habit of doing things I didn't want to do, and anybody who expected me to explain myself at this point was too late to the party to get their wish.

I didn't know if Tonya would be annoyed about the lull or not, but after my conversation with Hector about the AR-15s, I felt like celebrating. It was good news for once. I debated it, but finally texted her around seven. She'd often accused me of only wanting to see her late at night, which had been the truth sometimes, so I figured the early hour was itself a kind of apology.

*Sup?* I texted her.

She didn't text back for forty minutes, though I could see she'd read the text about ten seconds after I sent it.

*Meeting a client*, she said at last. *Then to see my weed guy. Need any?*

*Sure. I'll take a half. I'll pay you back.*

*Cash app me 200.*

*WTF? I can get a whole ounce for 250.*

*Then call your guy. That's what mine charges and he always has good shit.*

Fucking Cash App. The only people who used it were hookers and dealers. I sent her the money.

*Done*, I texted. *Highway robbery.*

*Ok see you after.*

*Do me a favor and stop by the grocery store on your way?*

I've found that when women have already decided you're an asshole, it's good to push the envelope a little, let them be reassured of their appraisal of you. Women deal with assholes all the time, so it puts them at ease when you act like one.

*Ugh. Text me a list and send the money. I'm walking into a hotel room.*

Tonya finally showed up at around nine dressed in a pair of tight black jeans and an AC/DC T-shirt.

"What do you know about AC/DC?" I said, taking the grocery sack from her arms and leading her to the kitchen. I was already pretty stoned, and I'd drunk most of a bottle of wine.

"I know my eleventh-grade algebra teacher liked to fuck to *Thunderstruck*," she said. "Is there more?"

"Jesus Christ." There was, in fact, a great deal she should know about AC/DC, but I figured it might be best to avoid starting the evening with a lecture, so I left it at that.

Tonya took a glass from the cabinet and helped herself to the last of the cabernet I'd opened earlier.

She made a face. "This wine is awful."

"Is it?" I started unloading the groceries. Eggs, butter, cookie mix. "You really fucked your math teacher?"

"You don't listen to anything I say, do you? I told you I didn't lose my virginity until I was twenty-two. But I did have a crush on him."

"I listen. I just forget."

Tonya pulled the baggie of weed from her back pocket and tossed it onto the counter beside the ingredients.

"Why the sudden urge for edibles?"

I opened the bag and inhaled.

"They're for a friend."

"I didn't know you had any friends other than Luis."

"They're for Luis."

"Ah, of course. Have you made edibles before?"

"Not since college."

"Well, step aside," she said. "I did some Googling at the red lights between the store and here."

While Tonya set up a double boiler, following the directions she'd found, I opened another bottle of cab and filled a new glass for her.

"See if you like this better," I said.

She took a sip and nodded her approval. "Much better."

I leaned against the counter watching her at the stove. She really did have an incredible ass. I could watch it all night.

She turned and caught me looking. "I know I've gained weight. You don't have to tell me."

"Yeah, right. You look incredible, and you know it. Enjoy it. Women age faster than men, you know."

"What a shitty thing to say, you asshole. You think you're aging so well?"

By her expression and the way she swiveled to look at me, I could tell she was about to go on a real tear.

"I'm sorry, I'm sorry." It was a flashback to thousands of similar occasions throughout my life. "I was only kidding, I swear. You're beautiful. You're the most beautiful woman I've ever been with in my life."

She turned back to the stove. "I know I am. But you're still an asshole. Tell me what happened this week. While you were ignoring me. Did bringing the table downstairs help at all?"

"It did, actually. You might call it the first pebble in a landslide."

"That's a good thing?"

"Seems like."

I told her everything that had happened, knocking out three complete AR-15s, hanging out at the hospital, the conversations with Hector. I didn't mention the drug run. Weed and psychedelics were fine by Tonya, but I always felt a cloud of judgment surrounding my coke use. I kept the updates strictly gun related.

"Is Hector going to be your first client?"

"My first buyer. In this field we call them buyers. We're not quite there, but I can tell he's interested."

"Great. So, what's left to do?"

"Kind of a lot. I still have to convert an AR to full-auto to sell him. I have to track down a router I know is somewhere around here. And a router bit. And I have to find this one jig somewhere up in the gun room. Then I have to clear off a work surface in the garage to do the actual milling. And I have to grind off all the serial numbers. The work itself I can do in an hour or so. Probably faster, after I get in the groove."

"Cool. While the cookies bake, we can find all the shit you need, then I can take a hit of this weed, and we can watch some TV and chill."

"Helping me with the cookies is more than enough, honey. We don't have to do any of this shit tonight. I can do it tomorrow."

"I bet you told yourself the same thing yesterday."

As always, she was right.

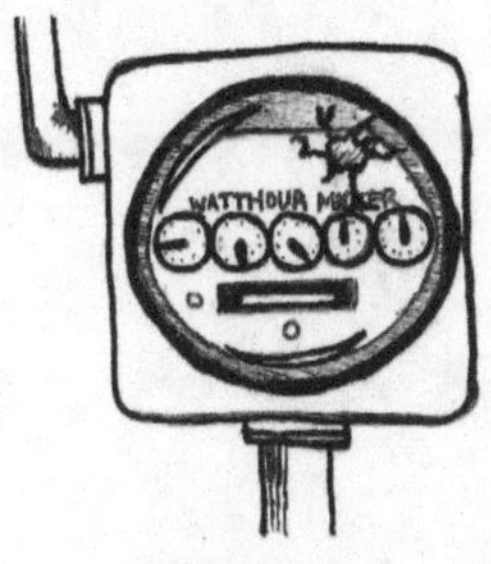


# THE ELECTRIC METER TRICK

I remember helping my grandfather with the electric meter trick. Believe it or not, before everything was digital, the turning wheel of a meter mounted on the outside of your house measured the electricity used on your property. Someone from the city came out once a month to mark down the numbers, and you were billed accordingly.

My grandfather drilled a little hole in the plastic cover of the meter.

"You got that BB I told you to get?"

I handed it to him, and he pushed it against the hole until it popped through and settled at the bottom of the meter.

"Now, when the guy checks, it'll look like someone shot it with a BB gun. Hand me that wire hanger."

He straightened the hanger and stuck it through the hole, maneuvering until it came to rest on the disk inside the meter, stopping it from turning. To the electric company, it would look like he was using practically no electricity.

Each month, he removed the hanger the day before the inspector came.

# 12

AFTER FIVE DAYS in the hospital, Luis was on the verge of breaking his all-time record for sobriety. He grabbed the bag of cookies before I could finish explaining what they were. We each downed two and sat watching TV and talking about the nurses while we waited for the high to come on. Hector arrived as we started to feel something.

"Quick, eat one," Luis told him. "They hit pretty fast."

Hector looked at the bag of cookies in his father's hand. "How strong is this shit?" he said.

"They're not gonna be that strong," Luis said. "Winston made them himself with his girlfriend. It ain't some kind of laboratory shit."

"Nah, I'm good. Next time."

I liked Hector, but it was a minor buzzkill having him there sober while the two of us were getting more and more fucked up. The three of us sat visiting in what would have looked like a pretty normal manner to someone on the outside, but it seemed to me—maybe because of the weed—our conversation was a stilted version of the way we'd spoken the day before. I got the feeling we were merely going through the preliminaries before some other, real conversation began. But I couldn't be sure, because with every passing second, I could feel the effects of the edible sweeping through me, amplifying in waves, filling my brain like a car filling with lake water.

After a few minutes, Hector leaned back in his chair, folded his hands across his chest, and said, "So, what makes you such an expert on guns?"

The question came out of left field. Or had it? I couldn't remember if we had even mentioned guns.

"When did I say I was an expert?"

"Oh, shut the fuck up," Luis said. "This motherfucker knows everything. He talks more about guns than he does about pussy."

"So, how did you learn it all then?"

"Just figured it out, I guess. A lot of it came from my grandfather."

I felt as if we weren't really in a hospital room at all, but on the set of a TV show, a world of props and scripts. Even the lighting seemed fake.

"Your grandfather taught you how to make silencers and machine guns?"

"No, he wasn't into shit like that. He taught me how guns work, though. Then I kept learning more as I got older. It's just how my brain works."

"What do you mean?"

"Well, like my second ex-wife, the whore, would walk into a room and hit the light switch, and it never occurred to her to think even for a second about where the power was coming from. To her, it was this thing that just worked. You know, just because it was supposed to. She never wondered about how it got from the world into her fucking hairdryer. But for me, I hit the switch, and for a second, I see like the Hoover Dam or something, and I see like turbines and generators and transformers, and then the ceiling fan starts to turn, you know? Same thing with the plumbing, right? My wife thought you flushed the toilet, the shit goes away. She never thought about where the water came from or where her shit went. It was just magically gone."

"Man, you're fucking high," Luis said. "Nobody thinks about the fucking Hoover Dam when they turn on the light."

"I know what he means," Hector said. "But what about guns?"

What did I mean about guns?

"I don't know," I said.

"This motherfucker is stoned," Luis said. "He don't know what he's talking about."

Luis started to laugh, and his laughter made me begin to laugh too, though I was trying to keep my head on straight because I understood what Hector was asking could be important.

"I think what I was saying about the Hoover Dam is, I used to think everybody thought that way. Like when I think about something, I see all the parts of it and how they fit together. So, it's that way for guns, too. Once I saw how they worked, it clicked in my brain."

Hector smiled. "I got an autistic cousin who says shit like that."

"He's not autistic, he's retarded," Luis said.

I didn't know if he meant Hector's cousin or me.

"It's like you being good at moving coke," I said. "It's how our brains work, right?"

"Yeah, maybe so," Hector said. "So, what do you carry?"

"Glock 19 usually. I like carrying a 43X better, but my dad got me a good holster for the 19 last Christmas, so... What about you?"

"Smith .357 in the Bronco. But I usually walk around with a Glock 35."

"Yeah, I had a few of those," I said. "They're big. Not that I mind that. But eventually I decided to move to mostly 9mm so all my ammunition's the same. That way, shit doesn't get confusing at the wrong moment."

A nurse walked in and addressed Luis.

"Two men who say they're your sons are in the lobby. You can only have two visitors at a time. What should I tell them?"

As far as I knew, it was the first time Manny and Chucho had bothered coming to the hospital since the day Luis was admitted. Hector and I stood to go.

"Still with these fucking Covid rules, man," Luis said as I nodded goodbye. "You're coming tomorrow, right?"

"Yeah, don't let those guys eat all your cookies."

"Fuck no, I'm hiding these things."

Hector gave his father a hug, and we pulled on our masks and left together.

In the hallway, we ran into Manny and Chucho, both of them wearing the blue surgical masks the hospital handed out at the door. Manny looked even thinner than the last time I'd seen him. He wore a D.A.R.E. T-shirt and a pair of filthy blue jeans held up by a belt that probably fit once but now only functioned because of an extra hole he'd punched in the leather. Chucho, as always, looked like a little sphere of a dude, at about 5'1".

"Hey," Hector said to his half-brothers.

"How's he doing in there?" Chucho said. "Is he awake and everything?"

"He's been awake for days," Hector said.

"Oh, cool."

"Hey, Wince," Manny said, "you still get Oxy for your back thing?"

"No. Like the last ten times you asked."

In the elevator, Hector said, "Fucking losers. That reminds me, I got a date on the junkyard thing. This Thursday, seven o'clock. You still in?"

"Absolutely."

# 13

WHEN MY MOTHER called at eight thirty in the morning, I answered only because I knew how shocked she'd be to hear my voice at that hour. I'd been up since seven, and though I hadn't changed from my robe, I had managed to clear a surface on the workbench in my garage. She called as I was going inside to get dressed.

"Well, you're up early," she said.

"Just because I don't answer the phone doesn't mean I'm sleeping."

"I know that. You wouldn't believe the morning I've had. Your father tracked mud into the laundry room, so I finally finished cleaning up that mess."

"I'm sure he got an earful."

"Is your friend still in the hospital?"

"Yeah, he's still there. Maybe another week they're saying."

"The last time your father was in the hospital, it was constant interruptions. You can't get any rest in there. Are any of his doctors white?"

"One of them, I think."

"When your father was in the hospital, there wasn't a single white doctor. They're all Indian now. Or Chinese. You can't understand a word they say. Did you show him that article I sent you about vitamin D?"

"Not yet."

Every week, my mother flooded my inbox with news articles culled from her Facebook feed.

"Well, make sure to show him. You have to order it from this one place or it does more harm than good."

"Hey, I can't talk long," I said. "Have you spoken to Sam? If you're really not going to give her that engagement ring, you should tell her."

"What I don't understand is why Daniel isn't the one calling me. Isn't it his job to get the ring? I've never heard of this thing where the girl arranges to get the ring and then gives it to her boyfriend and waits for him to propose. What if they break up and he keeps the ring?"

"I think we have to assume that's not going to happen."

"I still find it extremely odd. I mentioned it to the women in my Bible study, and they all think it's odd, too."

"People probably thought it was odd when you got knocked up and married at fifteen."

"Well, that was another situation entirely," she said. "Anyway, what do we even know about this man? I've only met him twice. I don't know a thing about Venezuelans, do you?"

"They've been together for three years, and it's Sam's choice, so I don't see what difference it makes how many times you've met him. I know as much about Venezuelans as you do. Anyway, his parents are Venezuelans, not him. The point is if you want to break her heart and go back on your word, you should tell her that's your decision, so we can get on with the drama."

By the time I was dressed in a pair of cargo shorts and an old T-shirt, she agreed to call Sam and send the ring.

"Now I have to figure out how to pack it. I'll have to go to the UPS store and have them insure it. I wonder if I should have it appraised first."

"Up to you. Call Sam and tell her. And please don't be negative. She's got enough going on."

"Pretty rich coming from you."

I anticipated about an hour of work to convert the first AR-15 to full auto, but I ended up doing the whole thing in stages: tracking down shit buried in the wasteland of the gun room and searching for a spring I lost my grip on and sent sailing across the room. It was a pain in the ass, but it felt good to be working.

Luis texted at around six to ask when he should expect me, and I told him I wasn't going to make it. As always, he assumed it had something to do with Tonya.

"Why don't you bring her up here to see me? You afraid I'll steal her?"

My phone rang an hour later. It was Hector.

"Hey, man. You ignoring my texts?" He sounded different, a veneer of fake calm masking something.

"I didn't hear my phone. I'm working on something. What's up?"

"I'm at my dad's house, and there's a bunch of shit missing."

"Like what?"

"Like all his shit. You been over here?"

"Not since the day I found him."

"Well, somebody took his shit. I'm over here feeding the cats, and everything's gone. Manny and Chucho are the only people who've been here besides you, and they say they didn't take anything. There's some cash on the coffee table, but everything else—"

"Oh fuck, man, I totally forgot. It's in a suitcase behind the shed. I thought the cops might come when I called 9-1-1, so I put it all in a bag and stashed it in the backyard."

I heard Hector open the sliding door and walk across the weedy grass. "Yeah, there's a bag here."

I listened to him carry it inside and unzip it. After a long pause, he took a deep breath and said, "Cool. Alright then. Glad it was that simple. Good looking out."

"We good?"

"Yeah, man. Sorry, I was tripping out there for a second. Tomorrow still good for you? Six o'clock at my dad's place?"

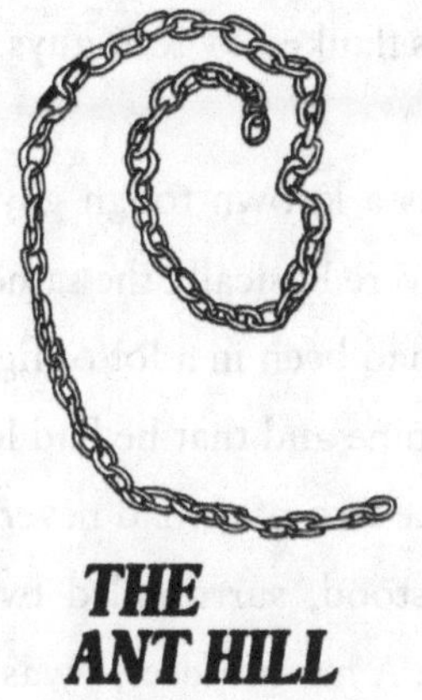

# THE ANT HILL

I remember at my school there were fights all the time, mostly down racial lines. If you went to a convenience store late at night and ran into a crew of Mexicans, you were going to get your ass kicked. I was there one night when a guy got killed that way, talking some idle shit to the wrong guys.

In tenth grade P.E., I stood in line with a bunch of other guys in the field behind the school. The coach was timing us doing 100-yard dashes, sending us out in pairs.

I stood in the scorching heat watching the guys ahead of me take their turns. I liked running and was good at it. I wanted to beat everyone's time. I had this fantasy the coach would ask me to join the track team, and I'd tell him to eat shit. That's what I was thinking about when Carlos Frias, behind me in line, picked up a small branch, stirred it around in a fire ant mound, opened the back of my shirt and shoved it in. Before I knew what happened, I was getting bitten. My back was scratched and covered in ants, the branch still inside my shirt. I reached back and pulled it out, scratching myself, pissing off the ants even more. I pulled my shirt off slapping at them, then turned to confront Carlos. Bastard was flanked by four guys in his crew, all of them laughing at me.

Carlos Frias was a known tough guy, kind of the Mexican version of me. We were basically the same size and had the same build. Like me, he had been in a lot of fights. It was known that I had lost none of mine and that he had lost none of his. But for whatever reason, the two of us had never fought.

Now there he stood, surrounded by his asshole friends. I didn't say anything. A minute later, I was at the front of the line, my back still stinging. The coach yelled at me to put my shirt on. It was my turn to run the dash. I ran like shit and went to the back of the line, past the guys still laughing at me.

The next day, I passed Carlos in a breezeway at school. When I saw him walking towards me, I took a three-foot chain from my pocket, whipped it over my head, and caught him right across his smirking fucking face. The chain wrapped around his head and opened up a series of cuts across his cheek.

I was suspended for three days, and my father had to go to the office to collect the chain from the principal. When he got home, he gave it back to me. I wasn't in trouble.

# 14

HECTOR WALKED AROUND the Tahoe looking for anything that might get us pulled over. I didn't know if this inspection was really part of his routine, or if it was some kind of signal to me that I was now seeing another side of him. He wasn't my stoner friend's kid anymore. He was a professional. Whether it was for show or not, I was a little impressed. I've long thought of this new breed of drug traffickers as lazy, stupid, and careless. It was interesting to find myself proven wrong by one of Luis's sons, of all people.

"This tint legal?" he said. "It looks dark."

"It's borderline, but it's legal. No one's ever hassled me about it."

"How long you had it?"

"Three years."

"You been pulled over in that time?"

"Yeah, couple of times. Speeding. Talked my way out of both tickets."

Hector nodded.

"How much are we picking up?" I said.

"Half kilo."

He stood behind my car and had me press the brakes. He looked at the registration sticker to confirm it wasn't out of date.

"If this whole cocaine dealer thing falls through, you could get a job doing annual inspections," I said.

"Alright, man," he said. "Looks good."

We were scheduled to arrive at seven. Even at that hour, there'd be traffic before we got out of the city, but it was still a relief when we finally hit the road. I've always felt comfortable behind the wheel of any vehicle. Whether it's a boat, a plane, or a fucking Panzer tank, I can drive it.

Hector was quiet as we left the city, texting and looking at a maps app.

Right after we passed the Austin city limits sign, apropos of nothing, he said, "You ever had to use your gun?"

"Yeah. You?"

"Yeah."

That was the end of that.

When the conversation resumed, we talked about Luis's condition and about the roads, how the city had changed in the years we'd both been there. He seemed so calm, I wondered what he had taken. I had debated taking a Xanax before leaving the house, but decided I didn't want to dull my senses. I'd done a couple of lines instead.

I asked Hector about his grandfather, who I knew had basically raised him. I was curious if he'd had learned about the coke trade from him. It did seem to be a family business.

"Nah, man. He's a ranch manager. Worked for the same family since before I was born."

"That's right. Luis told me you grew up on a ranch. I guess that means you actually know how to do things."

"Man, I know how to do everything."

"You ride a horse?"

Hector laughed. "Of course. My big dream is to buy that ranch for my grandfather. Move him into the big house and hire him a ranch manager. Let him be the boss for a while."

I liked those kinds of dreams. I'd had my share of them over the years, saving the people who had saved me, seeing the little guy get his due.

We were quiet for a while after that. I was thinking about the junkyard, trying to imagine what to expect. They're all a little different, but, as far as I can tell, I've never been to a junkyard that doesn't serve as a front for some kind of criminal enterprise. And I'd been to plenty. When I was young and getting into cars, it was because my friends and I were always tracking down parts. As an adult, I went for other reasons.

Junkyards exist in my mind as a perpetual cliché synonymous with drugs and violence and rednecks with wads of greasy cash. Being so close to one, passing through the Hill Country, I felt that old muscle tingling, a sensation that intensified many times over when I finally slowed and turned onto the gravel drive of our destination in the middle of nowhere. Before me, impeding my progress into the yard, stood a corrugated metal gate about the size and shape of a two-car garage door.

I stopped the Tahoe. To our left stood a small, brown outbuilding elevated five or six feet above the driveway with two men stationed on its narrow porch. The way the floodlights lit them from behind, they looked exactly like shooting targets, nothing more than vaguely threatening silhouettes until one of them moved.

"Flash your lights," Hector said. "When he comes to the window, tell him we're here to see Alfonso."

I flashed and rolled down my window.

The person who came down from the porch was a skinny Mexican kid in his early twenties with a patchy beard and oily black hair almost to his shoulders. I looked down at his holster, and what I saw there turned my stomach. One of the dumbest, most undignified firearms imaginable. This man, who undoubtedly prided himself on being a criminal and a gangster, wore on his hip a Glock 17 that appeared to have been dipped in green glitter fingernail polish. If Dorothy from *The Wizard of* fucking *Oz* had carried a Glock in her picnic basket, it would have been that one. Anyone who would wear such a gun—who was not a girl under the age of sixteen—could have none of my respect. I sneered at the kid when he got close.

"We're here to see Alfonso."

My tone gave the kid pause, but he only looked at me for a second before stepping back and raising his arm in the direction of the man still on the porch, who leaned inside the building's doorway and hit a button that initiated the slow, grinding motion of the gate. Gradually, like the pulling back of a curtain on the world's shittiest gameshow, the junkyard was revealed, along with another armed Mexican who waved us inside. As I pulled through the gate and alongside him, I saw the AR-15 he wore on a sling across his body was decorated in the colors of the

Mexican flag, the upper receiver green and the lower receiver and magazine red and white. Cartel guys. You gotta wonder what the hell they're doing on this side of the border if they like Mexico so goddamn much.

The man with the Mexican flag AR had a pretty impressive scar splitting his right eyebrow in two, and his accent was so thick you could hardly understand him. Luckily, he only had one line to speak. "Go till this ends."

I looked at him another moment before proceeding. For better or worse, I thought, these were my people. Not my ex-business partner Bryce or any of my wives or even my parents. I felt, as the junkyard pulled me into its belly, that I was in my element for the first time in years. I felt my shoulders drop and my chest loosen. I hadn't needed that Xanax after all.

I kept searching for the next cartel guy as we moved along, swallowed by stacks of burnt-out cars. Every few feet, I revised my plan for escape. I could turn around here, throw it into reverse there, crash into that pile and duck between those towers of Corollas, Tercels, and Accords.

The routes that led off to different parts of the yard had been blocked by wrecked cars dragged into place sometime before our arrival. Though I found myself creeping around several mazelike turns, there was only one direction to go. It seemed a safe bet they hadn't done all this for Hector and his half kilo. Presumably other customers had arrived before us and would arrive after us.

"Almost there," Hector said. "Two more guys up here in a minute."

So there were. Two men stood next to the steps of a single-wide trailer. One of them gestured for us to pull forward until he said stop. Hector unbuckled his seatbelt. He had already explained what would happen. I would sit waiting with the engine turned off being watched by the two guards holding AR-15s, and he would go inside and buy the coke.

Unruffled, Hector got out without saying a word and stopped at the front bumper where one of the two men frisked him. He wasn't allowed to take anything inside except the money, which he carried in an envelope in his back pocket. No one spoke. With my window down, I could hear their hands against the fabric of his starched jeans.

After they patted him down, one of the men led him up the wooden steps to the single-wide and opened the door. Hector disappeared inside. The man descended the steps, and he and his partner stood staring at me, sizing me up through the windshield—an old white guy with a pistol wedged beneath his thigh.

By then, I already halfway knew I was going to rob the junkyard. The passenger side headlight flickered and went out.

# 15

I MOTIONED TO one of the guards. As he approached, I could see he was another man in his late twenties, Mexican, of course, wearing jeans and a Western-style shirt. In addition to the AR-15 in his hands—it was black, thank God—he wore an old-fashioned shoulder holster made of brown leather. In the dim light, it was hard to be sure what he was carrying. It was a revolver. A Colt, I thought.

His face was expressionless.

"My headlight went out," I said when he got close enough.

"Yeah," he said.

"If you'll lend me a toolbox, I could probably find one that'll fit."

"Find one where?"

"It's a junkyard, isn't it?"

"You stay here," the man said.

"I'd hate to have to drive back to the city with a dead light."

He was already walking away. He went over to his friend and they both lit cigarettes. They looked at the Tahoe, and the guy with the holster said something to his friend that made them both laugh.

I slipped my phone from my pocket, opened Google Maps, identified my location, and took a screenshot. I zoomed out and took another. As I was putting the phone back in my pocket, the door to the single-wide opened, and Hector came out carrying a small, black bag about the size and shape of a camera satchel. He looked at the two men, and one of them dropped his cigarette and went up to the door. He stuck his head inside, received some kind of signal that all was well, and nodded to Hector that he was okay to go.

The other guard took out a cell phone and called someone, the guys at the front gate, I assumed.

Hector got in beside me. As I turned the Tahoe around, he opened the glove box and popped the hinge, revealing the void behind it. He pushed the bag inside.

"Headlight's out," he said.

"I know. Fucking thing. It won't be a problem. How'd everything go in there?"

"Smooth. It's just him and another guy. Hand them the money, they hand you the shit. Not a lot of small talk."

"They got a pile of coke or what?"

"In the other room, probably. Mine was waiting on the table. That's how they run it. You place your order in advance."

I pictured the room. A broken block of coke in plastic on the table.[1]

When we got back to the front gate, the guy with the Mexican flag AR was holding a phone to his ear. He said something into it, and the gate began to open. As we waited, I looked at the cars parked behind the building at the entrance: a new BMW 5-Series, wrapped in some kind of flashy iridescent bullshit, and two Chrysler 300s, one black, one silver. I never understood what it was about drug dealers and Chrysler 300s. I'd never even been in one, but there must have been something about them, because they were always sort of around.

As I drove through the gate, I looked back to see the two guys standing on the porch, gave them a wave, and turned the Tahoe onto the road.

"We came in the other way," Hector said.

"Yeah, maybe I'm paranoid, but I always like to leave a different way than I came."

---

[1] I could break it apart in my head. A half kilo is 500 grams. There are twenty-eight grams in an ounce, so you're looking at just under eighteen ounces total. Grams were what Luis sold, usually for a hundred dollars apiece, but Hector was more of an ounce guy. Say you get an ounce for eight hundred dollars, which is a great price but not unheard of. You could flip it whole for thirteen to fifteen hundred and make five to seven hundred without doing any real work. Or you could break it down into grams. Twenty-eight grams at a hundred each is twenty-eight hundred gross. After cost, that's about two grand profit for one ounce. Now start cutting it. Cut it by forty percent and you're suddenly selling close to forty grams. At that point you're north of three thousand in profit on the same eight-hundred-dollar buy. I never stepped on my shit that hard, but I knew plenty of guys who did and somehow never lost their customer base. Luis usually cut by around thirty percent, which still made a noticeable difference over time. I didn't know exactly what Hector paid for the half kilo, but I figured somewhere between fifteen and eighteen grand.

A black Escalade passed going the opposite direction, and I watched in my rearview mirror as it turned into the junkyard.

"Their next customer," I said.

Hector looked over his shoulder then glanced at the clock. "He's early. I think he usually spaces them out by like thirty minutes."

"What makes you think that?"

"He always tells me to come either on the hour or the half hour, and he tells me not to be late or early. One time there was some kind of delay, and he pushed me back to the next half hour. I think they're only out here like once a week, so they keep a pretty tight schedule."

"Wonder how late they stay."

"Eleven, I think. One time Alfonso said they don't like people leaving here too late at night."

I tried to get a sense of its scale as I drove alongside the junkyard. A high fence covered in corrugated metal ran along the road until it terminated at a dense wooded area. I guessed it was maybe five or six hundred feet deep.

Because I'd talked about it so recently, I guess, I noticed I was having one of those moments when things start coming together in my mind. There were all these pieces: the guys, the gate, the guns they carried, the weird path to the single-wide. Even with some of the parts missing, like the inside of the single-wide, I could see the way the whole thing fit together the way I could see electricity flowing from the Hoover Dam. I could even see how the junkyard was a piece of something bigger, a network that extended into Mexico.

When we hit the highway, our talk turned to different philosophies around cutting coke, a favorite conversation among traffickers.

"I used to cut it with inositol," I said, "then block it back up with acetone."

"Yeah," Hector said. "I used to do half and half. Cut half by maybe twenty percent, leave half untouched. Give people the choice. Tell them I got shit that's been stepped on a little, not much, or I have the shit that's untouched. Lots of people don't mind paying $80 for stepped-on shit if you're charging $120 for pure. Some people thought it hit even better with the inositol, which is probably bullshit, but whatever works for them, right?"

"You charge $120? Jesus."

"People know you have good shit, they'll pay. I don't sell in grams much anymore. I only keep them around as tips and for friends."

"And for personal use, I'm guessing."

"Nah, not so much anymore."

It was true I'd never once seen him use.

You heard it both ways. Some people said never trust a sober bartender. Other people said the opposite. Everyone I ever worked with partook, but Hector was different in a lot of ways than the guys I'd worked with.

When we got to Luis's house, Hector opened the camera bag in front of me. It looked the way I knew it would. A little half brick with the cartel's logo stamped into the coke—three crowns it looked like.

"Keep the half," I said. I didn't want Hector thinking I was some kind of junkie. "Give me five or six grams. A half ounce is more than I need. Anyway, you're going to help me with the other thing."

"That's right," he said. "Mr. Machine Gun. How about you start with getting me one? Three silencers, too, if you can handle it."

"That's no problem."

He didn't bring up the price, so I didn't either. I was still debating what to charge anyway.

My first order: one machine gun, three silencers. It was a start.

We ended up staying up late, bullshitting about the business and about guns.

I helped him weigh out Luis's grams.

"You don't want to cut it?" I said. "I can knock it out in no time."

"Nah, let dad's customers get pure shit for once. You ever heard of intermittent reinforcement? If you give people good shit once in a while, they keep coming back even though most of the time the shit they're getting is barely good enough."

"You just described all three of my marriages."

"Most addictive thing in the world, man."

# CABLE BOX HACKER

I remember meeting a guy at a party who worked for the cable company. Cable boxes had emerged pretty recently, and for a long time existed in the space between cutting-edge modernity and clunky mechanical devices. Before everything was digital, when your house received the signal, it received the whole signal, every possible channel. I knew there must be something inside each box that filtered out the ones you didn't pay for. I asked the guy, and he explained how a single microchip controlled the flow. He said it looked like a flat bug with twenty legs, ten on each side. To make a cable box get all the channels, all you had to do was bend back half of the legs so they didn't connect. I asked him which ones, and he told me. It was a five-minute conversation, maybe less.

I went home, opened my parents' cable box, and found the microchip. With a pair of needle-nose pliers, I bent back the legs like he said. Five on one side, five on the other. I plugged it in. Suddenly, we had all the channels. When I showed my father, he told his friends, and everyone we knew brought their cable boxes over so I could fix them for seventy-five dollars apiece.

## 16

BY THE TIME I got home from my night with Hector, it was after two. I did a bump off the kitchen counter, then thought better of it and did a rail. I went to the room I once referred to as my office, which contained an overstuffed filing cabinet and a huge, ugly desk piled with junk from my old work life. I hated the office even more than the upstairs. In that room, more than any other, I found myself remembering I hadn't filed my taxes in three years, a thought I could keep at bay in most other rooms of the house, although it had been rising up more and more often lately.

I'd made so much money at Keystone I hated to think what I must owe. Whatever the total, it was far more than I had left after a year of coke and takeout. I'd been audited twice in my forties, and it took plenty of fancy footwork to get out of them basically unscathed. I didn't look forward to a third round of dummying up receipts, airline tickets, email confirmations, and credit card bills.

I searched the drawers in the desk until I found a legal pad and one of the pens I used to insist on having at the office—the Precise V7 in black. I took them to the living room sofa and laid them on my grandfather's table. I sat down, turned on the TV, and drew a bird's eye view of the junkyard: the little outbuilding at its entrance, the crooked path we drove, and the door to the building where Hector made the pickup.

I got my laptop and, using the image from my phone, found the junkyard in Google Maps. Once I switched to satellite view, I could see the whole place. It was a rectangle, packed with piles of tires and rows of cars. Around it was just vacant woods from the looks of it. The only thing missing from the images online was the single-wide trailer, which hadn't been there when the pictures were taken. But I could see where it belonged. I tore off the sheet and sketched it again, a little better this time.

I marked Xs where the two men had stood outside the building at the entrance, another for the guy inside the gate, then two Xs for the guys standing outside the trailer, and two more for the men inside. Seven guys. Seemed like major overkill for anything short of Pablo Escobar. But, then, you never knew. Guys like that traveled in packs.

I used to sketch things like that all the time back when I was doing construction. Even in wealth management, I used to draw these illustrations of the three buckets I recommended for people: investment accounts, annuities, and life insurance. I liked my drawing. It felt good remembering how things I drew looked different than the things anyone else drew.

My phone rang. One of those middle of the night calls. It's true, you'll never know how terrifying the feeling is if you don't have a kid. When I looked, it wasn't Sam, thank God. And it wasn't my mother. Still, it wasn't exactly a welcome sight. The name on the screen said Stein, Frank N., my attempt at making something funny of someone who wasn't. My former business partner, Bryce Winters.

Inexplicably, I answered.

"Yeah?"

"Winsson."

That was all I needed to hear to know how fucked up he was. That's what Bryce always did, slurring away the "t" in my name. I have to admit, I actually felt some shred of nostalgia hearing his voice that way. Though I had spent every night for a year imagining ways I could eviscerate him, had even called Ruck to see if someone we knew could take care of him, some part of me apparently still liked the guy.

"Well, you sound fucked up," I said.

"Yeah, look who's talking. Sounding pretty chipper for two in the morning."

"Just enjoying my two fucking years of rest and relaxation."

"Oh, yeah?" Bryce audibly inhaled a line of coke. "I keep looking out for a postcard. What you been up to? Going out every night? Traveling? Getting laid? Bet you've been doing all kinds of cool shit. Not like you'd waste the whole time sitting around getting drunk and high alone, right?"

I reached for a witty retort, but when I hesitated a beat too long, he said, "I can hear those rusty gears turning from here. I'm sure you'll think of something clever to say by the time we get together tomorrow."

"Tomorrow?"

"Yeah, I need you to swing by the office. We've run into problems with some old files. Something with this couple, these Korean people. Elaine was telling me about it. They opened an equities account and bought some insurance from us, but I can't see where we ever invested the cash. Or maybe we bought the wrong thing? Then there's some other stuff. Elaine was looking at it, and there's kind of a stack. She said we should call you. I thought maybe you'd consider—"

"Ten grand," I said.

"For what?"

"That's my fee, if I'm freelancing."

"Ten grand? You gotta be kidding."

"I'm hanging up, Bryce. See you in hell."

"Five grand. Two hours for five grand is a lot."

"You saying I'm not worth it?" I said. "I'm offended."

"Seven grand. I guess I could do that. But you're actually going to fix this stuff, right?"

"I'll do what I can. Have the cash. I'll be there at one. Have Elaine order my usual from Bogart. I'll eat while I work. If you're late, I'm leaving. And the price goes up for a second visit."

I ended the call before he could reply.

Fucking Bryce. I almost smiled.

I looked at my drawing of the junkyard. Sometimes it blew my mind thinking about all the shit we had to do to make a living.

# 17

THERE'S ALWAYS BEEN a criminal dimension to wealth management, despite the fifteen or sixteen governing bodies supposedly preventing it. Somehow, I always ended up meeting guys who didn't do everything strictly by the books, so, yeah, I understood from the start that Bryce and the other guys at Keystone weren't saints.

Personally, I'd always kept things above-board during my years in wealth management. Whatever I'd done in the past, it was with and against other crooks. I certainly wasn't going to fuck with some old guy's retirement account. Maybe I sold some annuities I might otherwise have passed on because they were giving away a cruise, something like that. But basically, and certainly by the standards of Keystone, I was squeaky clean.

When Covid hit, no one at the office took it seriously. Like everyone else in my Facebook feed, I figured the whole thing was being overplayed. I mean, we're talking about a fucking flu here. It's not like it could derail the whole planet, right? Then it did.

Some of the guys I knew managed to pivot quickly—the whole Zoom revolution—but at Keystone we were in-person, old school suit and tie guys great at reading people and closing them face-to-face. I tried the digital thing for a while, but it felt like a whole different job. It wasn't even fun. My assistant, Timothy, could hardly get anything done. He was constantly fucking with every tech hurdle known to man—my computer and the camera and the lights. I went from badass wealth manager to QVC presenter. Finally, I told Timothy we were done. No more online meetings. We weren't making any progress anyway. I kept telling him the whole Covid thing would peter out pretty soon.

For a while, I thought about taking a leave of absence, disappearing to some beach in Mexico, where I'd lie around getting stoned on local herb, overtipping masked Latinos bringing me drinks. If I'd had a wife or a girlfriend, I probably would have told her to pack a bag, and we would have done it.

Instead, I stuck around Austin, getting fucked up at night and going into the office later and later each day. Eventually, there was no point in going in at all. There was no business to do. Still, the same expenses remained: salaries, rent, phone lines. I was a partner by then, so I was expected to kick in my share each month whether we were making money or not. And I did, for four months with no income. By month five, like everyone else, I was starting to doubt the world would ever go back to normal.

Bryce had become the head of Keystone by then, acting like a bigshot, calling mandatory meetings all the time, bitching everyone out for not bringing in clients, then trying to gas us up with bullshit motivational speeches about how we were going to be the last men standing.

Before this, Bryce and I had been friends. We'd screwed hookers together, gone on a couple of trips. We'd downed gallons of Don Julio on his boat. Plus, we'd made a shitload of money together, which is its own kind of bonding experience.

Bryce's low-budget charm got a lot of traction with clients. I saw him lie into the phone so often, grinning like he did, nudging me with his elbow, his slicked back hair some shitty knockoff of Gordon Gekko's. You could convince yourself a guy like that would never do you wrong because you'd already seen the other side of it. He'd let you see the real him. But that's another kind of con.

I was hemorrhaging cash I couldn't afford to lose keeping Keystone afloat. All you heard about back then was countries shutting down and every city's downtown transforming into a wasteland. No one was going back to the office, now or ever. The mainstream media kept playing images of refrigerated trucks supposedly filled with dead bodies because the morgues were overpacked.

I told Bryce in his office. I said, "Look, I gotta bail. I can't afford this."

He wasn't happy. Not that I expected him to be. Me leaving meant he'd be paying a hell of a lot more of the office's expenses on his own. But he didn't try to persuade me or anything. He turned on the charm, acting like it was no big deal, like the two of us would grab a beer as soon as the bars reopened, if they ever did.

"Who knows," he said. "Maybe after all this blows over, you'll end up coming back to Keystone."

"Yeah, maybe so," I said. "We had a pretty good run, didn't we?"

He was so gracious I almost second-guessed myself. Maybe I'd take him a bottle of scotch. Buy him a Chinese handjob at a place I knew he liked.

The next morning, I woke up and I couldn't log into my work email. I drove to the office to get my stuff, and my badge didn't open the door. Back at my house, I checked my personal email and found one from Bryce with no text in the body, just a single attachment: the signed two-year noncompete. From then on, Bryce refused my calls. Girls with whom I'd flirted for years would hear my voice, hurry through some scripted bullshit about how busy Bryce was, and—click—they were gone.

The next day UPS delivered my personal items. It looked like he'd swept the surface of my desk into a big box. A wrinkled photo of Sam at sixteen astride my Harley, the picture frame shattered. A purple, ceramic cow she made in the third grade, broken into several pieces. My nameplate, "Winston Fisher, Jr." scratched all to hell. My address book, with the names and contact information for every client methodically redacted with a black marker. Fucking Bryce. It was exactly the kind of thing he'd boast to me about doing to someone else. Except now he was boasting to someone else about doing it to me.

For years, the staff at Keystone had only seen me in expensive, tailored suits. Ralph Lauren Purple Label. It was a big part of my identity back then, the way other people saw me and the way I saw myself. I liked knowing how my suits should be tailored, how they should lie, where the cuffs should hit my shoes. The morning after the trip to the junkyard, the thought of putting on a suit for Bryce or for anyone else seemed totally ridiculous. Besides, none of mine fit anymore. I put on a pair of shorts, a T-shirt, and a pair of flip-flops, did a bump off the kitchen counter, and headed out.

After the night I'd had at the junkyard, I would have been happy to sleep all day, but I made it to Keystone by a quarter after one, long enough for Bryce to have to wait for me a while, but not so long that my steak would be ruined.

Angela, a front desk girl who I'd always taken the time to flirt with, didn't recognize me as I walked past. It was the beard, I guess. She tried to keep me from the elevators saying, "Sir! Sir!"

When I turned and looked at her, she sank back into her chair. "Oh. Bryce said you were coming."

Traveling to the third floor, I looked at myself in the mirrored doors, licked a finger, and cleaned white powder from my nostril.

Bryce and Elaine sat waiting in the lobby of the office suite. Bryce wore a grey suit and a pink tie, and Elaine a grey and black striped dress that reminded me why, for years, I had wanted to fuck her. I recognized the activation of Bryce's bullshit persona the moment he spotted me. It was like watching someone flip the switch on a robot. They stood to greet me.

"Casual attire," Bryce said, bringing his hands together in a clap. "I love it. And the beard. You look like you're about to build a log cabin."

"Hi, Winston," Elaine said in her usual deadpan. "Good to see you."

"You, too." I directed all my attention to her, ignoring Bryce. "Where am I working?"

"Conference room."

"You get my steak?"

"It was waiting for you at one."

"You're an angel. What are we looking at today?"

"A bunch of problems. You'll see."

Even smelling Bryce's fucking Axe deodorant made my fist clench.

"Lead the way," I said.

Elaine led me to the conference room and shut the door. An open laptop sat beside a stack of files, and, beside that, my steak and mashed potatoes on a white plate.

"I'm not sure what he told you," she said. "There are multiple issues."

"Any on files I put together?"

"One of them is a referral you took, but they didn't invest anything until you were gone."

"Before we get started, do you know where my money is? Bryce was supposed to give me seven thousand in cash up front."

"He had it in his pocket. I'll go get it."

"Do me a favor and tell him he better not try to 1099 me."

While she was out, I went down the hall to microwave my steak in the kitchenette and paused at my old office. More than a year later, it hadn't yet been filled by anything other than file boxes. I stood in the door a moment, expecting to feel something, but the only things that bubbled up were bad memories.

When I returned to the conference room, Elaine handed me an envelope.

"Think I should count it?"

"I would," she said.

I tossed it down on the table next to my plate. While I ate, Elaine opened various files and explained the issues with them. There were a lot of problems, most of which had to do with orders they hadn't filled and instances in which they'd purchased the wrong policies on behalf of clients. General negligence stuff. As she spoke, I knew I had absolutely no clue how to help with any of the problems at hand.

"I'm sure I could fix at least some of this shit," I said. "But I'll need another three thousand dollars to do it."

"Three thousand dollars?"

"Ask Bryce, would you? I don't want to see his face. If I have to talk to him directly, it'll be four."

When Elaine left the room, I called Timothy, my assistant of six years who was saved in my phone under the name "Boy Wonder." He answered on the third ring.

"What's up?"

"You at work?"

"Yeah."

Timothy, unburdened by a noncompete, had gone to work for one of our competitors the moment I left Keystone. We had stayed in touch, and it was understood that wherever I landed, he would follow. The place he ended up was way beneath Timothy, though he seemed happy enough when I checked in on him. He was twenty-five, nerdy as hell, and probably the smartest person I had ever met, at least when it came to that kind of shit. He had once confided to me that he had fucked only one girl his whole life, which surprised me, because I figured the number was closer to zero.

"If I gave you three grand, would you come to Keystone real quick and help me sort out a few things? Looks like Bryce fucked up a bunch of stuff, surprise, surprise."

"Is this the stuff Elaine called me about last week?"

"Probably."

"I told them I wasn't coming back after the way Bryce treated you."

"Well, now he's treating me to lunch and some cash."

"You guys make up?"

"Fuck no. But I got seven thousand for me and three for you. Can you fake a stomach ache and get here within half an hour? I don't want to have to come back here tomorrow."

"Yeah, I'll be there," he said. "Make sure Angela at the front desk knows. I don't want to be told I'm barred from the premises again."

By the time he showed up, I had the folders all laid out for him. There were six files in total. Beside them sat another envelope with Timothy's name on it containing three thousand dollars from Bryce's office safe. Timothy arrived wearing a pair of khaki pants, a tucked-in green polo shirt, a navy-blue backpack, and the same bushy, curly haircut he'd been wearing since the day I met him. He looked like a college freshman who'd just passed the Series 7. It was good to see him. I loved the guy.

He looked over the files. "When are we supposed to do all this?"

"We don't have to do all of it. Some of it. Say, three files worth. Like I said, I don't want to come back. We've got Elaine and whoever else is in the office at our disposal. Can we do it today?"

"It's after two o'clock."

"So?"

"The market closes at three."

This had been Timothy's entire career. Right out of college he had come to work for Keystone as an intern, then become my assistant. He was the only person on staff who could wrangle me or figure out how to solve whatever problems I created. When I made unkeepable promises to clients, Tim found ways to keep them. This was his junkyard.

"What can I get you to drink?"

"Water," he said.

Even Tim's drinks were square. I went down the hall to fetch him a bottle of Evian. For the next three hours, I got him whatever he needed as he solved each of the six issues at hand. I tried stopping him after a while, but he kept going. Some of the clients he remembered. He didn't like the idea of them getting screwed by Bryce. He could never stand a mess. No matter how many dirty dishes you left lying around, he'd clean them up if he went into the break room and found them in the sink. I couldn't tell you what he did that day, but the kid was a whiz on the computer.

"No matter what the new place is paying you, it's not enough," I told him in the elevator on our way out.

"Sixty-five," he said. "That's what they're paying me."

"Fuck's sake, that's terrible."

The last year he worked with me, I'd paid him that much in bonuses.

"Only ten months left in your noncompete, right? Then we can start our own thing or go somewhere else."

It was true, this had been the plan. We would start something new together or go to another office. We'd make so much money for them, they'd leave us alone, and we could do what we were good at. I would be the salesy chameleon rainmaker, and Tim would do everything else. It had once sounded like a dream.

"Let me see your envelope," I said.

He handed it to me.

I took mine from my pocket and gave it to him.

"Three for me, seven for you."

"You don't have to do that, Winston. I'm happy to help. Seriously."

"Take it. Call next week if you feel like catching up. I'll buy you lunch."

"You're gonna sober up a little before we start our new thing, right?"

"I haven't had a drink all day."

"Yeah, well, that's a lot of bathroom breaks to take in three hours. And you're looking a little perspired."

"You're always perspired when you've put on this much weight," I said. "Anyway, I've got it under control."

Every time I told someone that I had everything under control, I thought of my uncle who used to say, "I know *everything*." He was eventually killed when he fell off a tractor and was run over, but he had always made us laugh. I'd told Timothy I had things under control many, many times. In my defense, it turned out to be true practically every time I said it. But now that seemed like a long time ago, and as I said those words to Timothy in the sinking elevator at Keystone, I didn't feel quite as confident as I would have liked.

The doors opened, and we passed through the lobby. Heroically, I resisted the temptation to spit on that Angela bitch on the way out. I kept thinking I'd see Bryce somewhere along the way. Surely, he wouldn't entice me to Keystone and let me go that easily. I half expected him to pop out somewhere in the parking lot, but, as I sat in the Tahoe doing a bump off my thumb joint, he was nowhere in sight.

# 18

I DIDN'T FEEL much like going to the hospital, but a night at the junkyard and a day at the office was the most work I'd done in more than a year, and I knew if I went to my house, I'd go back to bed or end up sunk in the couch all night. It had been three days since I'd seen him, and Luis had texted asking if everything was alright.

The gay nurse was leaving Luis's room when I arrived. "Well, look who's here," he said. "I was just asking Mr. Rivera where his bodyguard had been hiding himself."

Gay guys are always like this with me, acting like I'm going to play Lou Grant to their Mary Richards. My second ex-wife, the whore, had a weekly night out with four or five of these types. I went out with them one time. The whole night, these guys thought it was such a trip to touch my leg and refer to me as "miss." Shit like, "Uh oh, looks like Ms. Fisher is getting upset."

Luis was in bed, looking better than when I had last seen him. His color was improved, and his stomach, which had grown into a paunch that put even mine to shame, seemed to have fallen like an undercooked cake.

"Hey, vato! You're here!"

It still surprised me that anyone was as excited to see me as Luis was. If we weren't real friends before, I guess saving his life had done it.

I sat down in the chair beside his bed.

"What are the doctors saying today?"

"Same old shit. Had dialysis all morning."

"How's that?"

"It's alright. Sitting around watching TV. How'd things go with Hector last night?"

I'd been debating how I wanted this conversation to go. "It was interesting," I said. "I haven't been to a junkyard in years."

"Did you see the guns those guys were carrying?"

"Yeah, you probably liked the Mexican flag one."

"I didn't see that one. I saw a pink one and like a green, glittery one. How did it feel going out and seeing the real cartel guys?"

"Looked like regular Mexicans to me."

"Did you have your gun in your hand?"

"On the seat."

"Me too, that's what I did."

When I didn't say anything else, Luis said, "Hector said you did real good. He said he could see you had experience. Most guys he don't trust out there."

"Yeah, it was fun, but I don't know that I'd want to do it again. Seems kinda risky."

Luis paused. "Oh, so you didn't like it then?"

"No, it was fine. I was glad to help. Plus, he gave me a few grams for my time. But I probably wouldn't want to make a habit of going out there. Lucky for me, you'll be home soon, so you can handle it again."

A look of disappointment crossed Luis's face. "Oh, so you don't want to go back? I thought maybe you and Hector could make a good team."

"Nah, I'm not going back to that. I've fantasized about it plenty, but last night was a good reminder that I'm probably not cut out for it anymore. Sam always says I'm a softie. Maybe she's right."

"Big old teddy bear, huh? That's too bad."

"Too bad why? You worried about Hector or something?"

"Not worried. But I figured he'd be better off with a guy like you having his back."

"You don't have anything to worry about. Hector's smart. He's got his shit together. He might even be smarter than I was at his age."

"Yeah, well, that ain't saying much."

"You polish off those cookies?"

"No, I still have some left," he said. "Why, you want one?"

I hesitated a second, then, like always, thought, "Fuck it."

After my first bite, I said, "We still got the guns, though."

"Who?"

"Me and Hector. We're still doing the gun thing. It's not like I'm reformed or anything."

Luis grinned, spilling crumbs from his mouth. "Oh, that's right. The guns. Right on."

# A TEXAN IN CALIFORNIA

I remember after rehab, I got kicked out of a halfway house in Phoenix when I got caught with a Colt .45 I'd bought off some junkie. I ended up in another halfway house in Los Angeles. I had the hots for this hippie chick from Narcotics Anonymous who wore practically no clothes. People back home didn't dress like that. Even people in the movies didn't. One night, she finally took me back to her place. It was a tiny apartment, and she was crashing on a sofa bed. Her roommate, who had the bedroom, was out of town. The two of us fucked, and she had these perfect teardrop tits I can still remember.

Sometime after we fell asleep, I awoke to someone messing with the doorknob. Before I knew it, I was on my feet, naked and standing behind the door holding the Colt.

A second later, the door opened, and the couple who entered found themselves looking down its barrel. There was a lot of commotion for a minute before I understood that the girl was the roommate. She hadn't left town after all, and instead, I guess, had gone out slutting around.

When everyone calmed down, the roommate and the guy went to her bedroom, and the chick I was with said, "I've never been so turned on in my entire life. Fuck me right this second."

She wasn't kidding. She was louder than before and didn't seem to care about the couple in the other room overhearing us. And of course, they weren't going to come out complaining after what they'd been through. The thing I kept thinking was, I guess they're not used to Texans out there.

# 19

I'D ALREADY BUILT three ARs, which was two more than I'd need to fill Hector's order. He had also asked for three silencers, so I knocked those out without much work.[1] The only question now was where I was going to test it all to make sure I hadn't fucked anything up. The first place that came to mind was the ranch of an old buddy, Garland.

Garland had actually been with me the first time I converted an AR-15 to full auto twenty years earlier. To try it out, the two of us drove to a spot I'd found not far from my house. I always thought it would be a good place to shoot, because it was cut off from the road by a running creek. I figured no cops would

---

[1] To say that I "made" the silencers is legally accurate, but technically a little exaggerated, since I didn't do much more than drill holes in an existing product. They're sold as "solvent traps." Picture something that looks exactly like a silencer and screws onto the barrel of your gun like a silencer. The idea is that you attach it to the barrel, spray the internals of the gun with cleaner, then it all runs into the trap, and you unscrew it without spilling a drop. They're pitched as some kind of ecological solution. As if, without a solvent trap, you'd obviously dump a bunch of grease into the grass. I mean, even if you were using it for its intended purpose, which no one does, you'd probably dump it in the grass anyway. It traps the solvent. It doesn't dispose of it. Sellers always cover their asses by saying they advise against making any modifications to their products, but at the end of the solvent trap, there's usually an indentation right in the center, and that's where you put the drill bit. Once the hole is drilled, you've got yourself a silencer.

trouble themselves enough to cross the water to check out whatever was going on.

We fired into the hillside like kids, passing the gun back and forth, setting up cans and targets to mow them down. It was kind of an incredible moment—an actual machine gun I'd made with my own hands.

When we stopped firing to reload, I heard tires on gravel and, though I couldn't see through the trees, I knew it was the cops. I ran to the truck, hid the modified AR under the backseat, grabbed two unaltered AR-15s I'd brought along and handed one to Garland.

"The cops are here. Shoot as fast as you can."

Garland didn't hesitate. We began firing one after another in perfect rhythm.

I had to hand it to them, the cops did cross the creek to check it out.

"Cease fire!" they yelled.

We were shooting in the opposite direction, both of us wearing ear protection. As they approached, we pretended not to hear them, changed magazines and knocked out another thirty rounds apiece. They were twenty feet behind us yelling at the top of their lungs when we were empty again. We did a pretty good job of acting startled. Garland even jumped and clutched his chest, the fucker.

The police had received reports of automatic gunfire in the area, they told us.

"Oh, man," I said. "It never occurred to me anyone would hear us out here. And anyway, we were just shooting at the same time. I guess I can see how someone might mistake that for the sound of a machine gun. I've never even seen one in real life. Have you, Garland?"

Of course he hadn't.

After that, we always shot at his ranch an hour outside of town.

Garland and I had the kind of friendship where we didn't need to check in with each other all the time. We hadn't spoken in about three years.

I called, and he picked up right away.

"Wince! You're alive. Candace mentioned your name not ten minutes ago."

"Yeah, my ears were burning. How you two doing?"

"Oh, we're hanging on. We're in Mexico. Both of us got fuckin' Covid."

"No shit, you guys okay?"

"I'm fine. It was like the flu. Better than taking some vaccine and dropping dead in six months. Candace still feels like shit, but it'll pass. We're not worried."

"Shit. Well, I'm glad you're okay. Damn. I was hoping you were in town. I need to try something out. Like the time those cops fucked with us, remember?"

"No shit? Wish I was there. Well, go use the ranch. Nathan's there with his girlfriend. I'll tell him you're coming. Hang on, I'll send you his number."

My phone pinged with the text.

"It's not a problem?"

"You kiddin?"

Later in the day, I called. I hadn't seen Nathan in years, but I'd heard about him cycling through a bunch of fuckup jobs in the city—roofing, cell phone sales, some other shit I can't remember—before finally moving back to his parents' ranch.

"Dad told me you'd be callin'" he said when he answered, using a good ol' boy accent thicker than I remembered. I could hear the wad of dip in his lip. "Come out any time. If me and Kelly are around and we hear shootin, we'll come out and say hi. The key's in the same place as always."

I told him to listen out for me tomorrow.

# 20

I HAD PINNED the junkyard map on the wall days earlier. It was such a vague illustration, it could have been anything. Of course, when Tonya showed up, she noticed it right away.

"What's with the drawing?"

She was always prying about every little change at my house. You'd think in the mess, she'd miss a few details.

"It's a sketch of the deck I'm planning to build."

"You're finally going to replace that thing? Thank God."

The back deck was in pretty rough shape and had been for a while. Two years earlier, I had accidentally burned down most of it with a stray cigarette cherry. I had paid someone to come demo and haul off the burnt section, planning to replace it. Of course, I'd never gotten around to it, and now what remained was a precarious, rotting platform with no railing.

Each morning, I stood out there taking a leak while I smoked, envisioning the elaborate deck I would one day build. I pictured cushioned sofas, an outdoor cooking area, and a sound system, so when my friends came over, we could jam out, smoke weed, drink, and reconnect.

Tonya must have sensed I was about to go into my usual daydream about the deck, because she cut me off at the pass.

"Where are you taking those?" she said, pointing at the bag of silencers I'd packed.

"I'm testing the AR and the silencers at my buddy's ranch tomorrow."

"You really have to test them?"

"Yeah."[1]

"You have to test the silencers, too?"

"Not really, but I figure as long as I'm going out, I might as well."

"Are they really that quiet?" she said.[2]

Silenced, it wasn't as loud as a regular discharge, but it still sounded like gunfire. Especially the first round.[3] The bark exploded from the tree where I shot it.

The sound still echoed as I unscrewed the silencer, and Tonya said, "That's enough. Seriously."

---

[1] Every gun that's ever sold has been tested first. It's called "proofing." Usually, it's done with a high-pressure cartridge to ensure the weapon won't fail under stress. What I was doing was different. What mattered to me was that when I popped in a magazine and pulled the trigger, bullets kept on coming.

[2] Suppressors aren't what people think they are. Even "silenced" weapons register as gunfire. In the UK they call them "moderators," which seems more honest. The sound of gunfire comes from two places: the gases being expelled from the explosion of gunpowder—think the popping of a champagne cork—and the sharp crack when the bullet breaks the sound barrier. The second is going to happen whether you've got a silencer or not unless you're using subsonic ammo in addition to a suppressor. But subsonic ammo is subsonic because it doesn't break the sound barrier, which means it's slower, usually heavier, and might not cause the kind of damage you want.

[3] The first shot through a suppressor is always loudest because oxygen in the suppressor ignites and makes a sound people call the first-round pop or FRP. Subsequent shots are quieter because the oxygen gets displaced by the gasses of the previous explosion. The ideal, if you really want to dampen the FRP is to fire first into something like a mattress.

I ignored her, and she covered her ears. I fired again at the tree. This time, the sound was enormous. Tonya jumped, and my ears began to ring.

We ducked inside as the lights came on in my neighbor's house.

"You're such an idiot," Tonya said.

"Just demonstrating."

"I'm done. That's enough excitement for me."

"Nothing's going to happen," I said. "It's nothing."

"Can you chill the fuck out once in a while?"

"Half the time you complain I'm not doing anything at all."

"If you can act normal for like five minutes, I'll get in the shower and hang out for a while. Otherwise, I'm leaving. Your call."

"You better go turn it on then. You know it takes forever to warm up."

"Then no more guns tonight, Winston. I mean it."

While Tonya washed up, I kept looking out the front door, and sure enough, through the leaded glass I watched two cop cars appear on the block. After cruising up and down a while, they stopped and spoke with my neighbor, who stood in his front yard in pajama pants, gesturing up the street. While they spoke, my cellphone rang, and I went to the couch to get it.

Stein, Frank N.

"Bryce." I took the phone back to the front of the house and watched one of the cops hand his card to my neighbor.

"Winsson."

I didn't say anything.

"Hey, man. I wanted to say thanks for getting that stuff worked out at the office. You and Timothy are a good team."

"Yeah, too bad we can't work together for another ten months."

"We should talk, man. Covid's over."

"Talk about what? We don't have anything to talk about unless you release me from the noncompete."

"You don't need to be released from anything if you come back to Keystone. Anyway, you're the one who signed it, Winsson. I don't know why you keep blaming me when—"

I hung up before he could finish.

When Tonya got out of the shower, I didn't mention the cops, but I did tell her about the call and bitched about Bryce. She listened for a while, drying her hair.

"You know, I wish you'd still fuck me once in a while," she said.

"Where'd that come from?"

Of course, I knew exactly where it had come from. Whether they admit it or not, all chicks are turned on by guns.

"Just a thought," she said. "Sorry to interrupt your boring rant."

"Yeah, well, I'd be down, but I definitely have a case of coke dick tonight."

"You still have those shots?"

For a while, when the pills stopped working, I'd used injections to get hard. The kind you stick straight into your cock that make it stand at attention whether it wants to or not. It freaked some girls out, but not Tonya.

"I haven't had any of those for a while," I said. "No insurance."

"Well, you still like dining at the Y, don't you?"

"I could do that," I said.

And, like I knew I would be, I was still pretty fucking good at it.

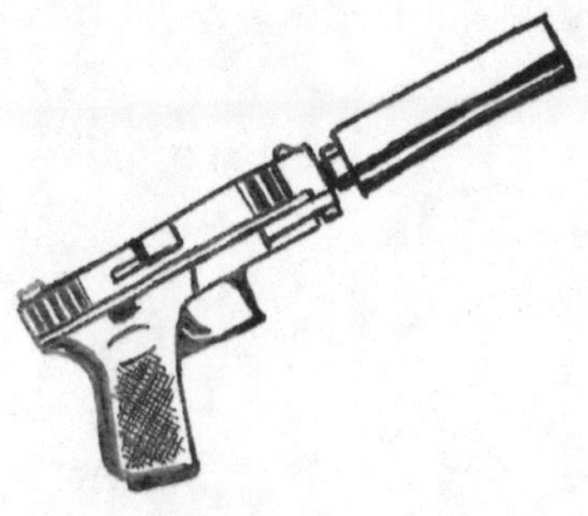

# OLD FAITHFUL

I remember the first time I got a silencer the legit way.[1] My form 4 took months to get approved. When it finally did, I went to the shop to get the silencer out of jail. I was so excited to pick it up, I had it screwed onto the barrel of a Glock before I even got into my driveway. I couldn't wait to try it out, so I rushed to the gate on the side of the house and fired a shot into the ground the second it was closed behind me. Water instantly spewed up from the hole and didn't stop.

---

[1] You can't just walk into a gun store and buy a silencer. That would be too easy. You have to spend somewhere between $500 and $1,200 for a decent one from a reliable brand. Then you have to submit a Form 4 with the ATF that they can take anywhere from a few weeks to a year to get approved. On top of that, you have to submit it with $200 for a so-called "tax stamp." A scam, sure, but if you want to shoot without wrecking your hearing you pay it.

The yard was probably fifty feet by a hundred feet. The water line was an inch wide, but somehow, I'd managed to hit it. Just my fucking luck. And the silencer wasn't even that quiet.

# 21

TONYA SPENT THE night. While I showered the next morning, she went out to the front porch and made a call. By the time I was dressed, she was looking through the fridge and coming up empty.

"Oh, well," she said. "We can get something on the way to the ranch."

I couldn't remember inviting her, but maybe I had. Having her along for the ride actually sounded nice.

My mother texted as I drove, and I handed the phone to Tonya.

"What's she saying?"

"Oh, God, it's a long one," she said. "Mom's on a tear."

Tonya had already heard plenty of bellyaching about my parents, but as we drove, she heard a little more. I told her all about the latest ring controversy brewing between Sam and my mother. The new problem was that Sam had received the ring, but when she opened the package, she discovered it didn't look at all like she'd remembered. The ring was old fashioned, and not in a way that was trendy. Sam had cooked up a new plan. She wanted to disassemble the ring and have a new one made using the stones from the old one. It was guaranteed to upset my mother and I had, once again, been enlisted to reason with her. As if such a thing were possible.

Tonya made some suggestions that probably would have been helpful if my family had been a more normal one. I kept saying, "You don't know how my mother is." I'd been repeating that since I was six years old.

We got to the ranch a little after noon, and I retrieved the gate key from an empty birdfeeder where Garland had hidden it for years.

I drove us across the property and could see the ranch house and Garland's trucks when I got past the trees. Last time I saw the hillside where we'd shot, it was a pile of brown dirt. Now it was covered with grass and saplings, and I almost drove past it. I parked thirty yards from the berm and took the AR from the backseat.

"God, it's peaceful," Tonya said. "Your friend Garland is a lucky guy."

"Muffs or plugs?" I said, pointing to my ear.

"Plugs."

I dug through my bag and handed her a little plastic case. She put her earplugs in, and I did the same.

I popped a magazine into the lower receiver, pulled the charging handle, and lifted the stock to the junction between my shoulder and chest. I flipped the selector switch to Auto.

"Ready?" I said.

If she answered, I didn't hear her. I waited a beat, then squeezed.

Empty cartridges roared out like a stream in my peripheral vision, the recoil into my shoulder fast and rhythmic. Saplings on the berm snapped and broke in half.

"You look good shooting like that," Tonya said. "Something about it looks right."

I offered her the gun. "Want to try?"

"No."

I blew through forty bucks' worth of ammunition before I was satisfied. Then I took out the silencers I'd made and tested them one at a time using the same Glock as the night before.

As I was finishing up, I heard engines. I turned to see a pair of four-wheelers headed our way.

"Should we hide this stuff?" Tonya said.

"Nah, they don't care."

When they got close, they circled us a few times, kicking up clods of dirt with their tires. Nathan looked fat. He'd put on thirty or forty pounds since the last time I'd seen him, and he now wore a patchy beard. Finally, they slowed to a stop, hopped off the four-wheelers, and Nathan came over to shake my hand. He was dressed in camo, smelled like body odor, and had a load of dip in his lip. Still, you could almost see the good-looking kid in there itching to get out.

"Look who grew a Covid beard, too," he said, using the same exaggerated accent I'd noticed on the phone. He slapped me on the shoulder like I was some twenty-two-year-old buddy of his.

The girl stepped forward and shook my hand in a pretty mannish way. "Kelly Jackson," she said.

"Winston."

"Good to meet you. Heard good things."

Kelly wore on one side of her waist a walkie-talkie clipped to her jeans. On the other hip, a holstered nine-millimeter Glock 17. Kelly wasn't ugly exactly. She was short and slightly pudgy, squeezed into a pair of jeans and a pearl snap shirt made for women of her build, her hair pulled back into a ponytail. I had fucked girls like that when I was his age. That kind of meat on her bones could be sexy in bed, but boy it spread like wildfire.

"This is my friend Tonya," I said, suddenly feeling a little nervous about the introduction.

Tonya seemed to take it in stride. "I love your ranch," she said. "It's so beautiful here."

Nathan looked her up and down. "Think so?"

"You should see it in the spring," I said. "Though it's been years since I've been here."

"Yeah, that's right," Nathan said. "We've gone full prepper since then. Got an underground shelter even."

"No, you had that last time I was here."

"No shit? Guess we ought to check expiration dates on rations sometime soon. Remind me about that, Kelly."

Kelly nodded, and I knew she would remind him.

"Me and Kelly are going to start building our house right over there next year," Nathan said. "We almost bought this one place over in town, but they got all these rules about what you can and can't do, and, hell, I'm twenty-three. Nobody who's not my daddy is going to boss me around on my own property."

"Plus," Kelly butted in, "they got a whole mess of Blacks in that neighborhood, and I told Nathan, I can deal with a bunch of rules if I have to, but you know even if we follow them, some other people won't, so forget it. Besides, country folk are better in the country, especially if you're trying to raise babies."

"Y'all thinking of starting a family?"

"I already got two," Kelly said. "They stay with my mama, but they'll move out here when the house is finished. Becka's seven, and she's already shot just about every gun we have."

"Good for Becka," I said. "My little Sam was shooting by the time she was five. She's a better shot than me any day."

"Oh, Becka ain't afraid of nothing."

It seemed to me that with every passing second their accents were unwinding into a kind of muddy sludge.

Finally, I couldn't resist looking at Tonya. Her face was bright red. I knew what a lefty she was from the hundreds of conversations we'd agreed not to have, but surely she'd heard all this shit before. It was Texas, after all.

"Y'all come on up to the house," Nathan said. "Got us some deer meat I know y'all'd like, and Kelly knows how to cook it, too."

Tonya narrowed her eyes at me.

"Thanks, but we better get going," I said. "I'll be back in the next week or so. I've got some other stuff I want to try out."

"Nice meeting you two," Tonya said. "Good luck with the house."

"Y'all come back anytime," Nathan said, spitting on the ground. He leered at Tonya. "Always liked getting to meet Winston's girlfriends. Each one's prettier than the last."

"Y'all be careful in the city," Kelly said.

"Jesus Christ, those two are horrible," Tonya said as we began our trip back through the property. "Are all your friends like that?"

"He's not my friend. He's my friend's kid."

"Well, I'm guessing he learned it somewhere. Is that how all your friends talk? Fucking paranoid, racist...Do you talk like that? If I hadn't been here, you'd be up at the house saying the n-word and making plans for the next Civil War?"

"Honey, stop it."

Just because I was friends with someone didn't mean I co-signed everything they said, let alone everything their kid's girlfriends said.

"Not all my friends are like that smelly kid," I said after miles of silence. "You can't get bent out of shape at me for something someone else said, alright?"

"Okay."

"I'm relieved the gun worked."

"Yeah, you were actually smiling for the second time in twenty-four hours."

Her phone rang on the way back, but she sent the calls to voicemail and replied by text. When I asked if she wanted to get anything else done while we were out, she said she was in a hurry to get back. The tension in the car was familiar from dozens of car rides from my past. I tried to make conversation, but nothing stuck.

When I pulled into the driveway, she didn't wait for the garage to finish opening. She grabbed her purse and hopped out.

"Thanks," she said. "I'll talk to you later."

She closed the door.

"Yeah, fuck you, too," I said under my breath.

# 22

I WOKE UP late the next afternoon to two texts from Sam, three from my mother, and one from Ruck, asking what I was doing for Thanksgiving. As I was smoking my first cigarette of the day on the front porch, a text came in from Hector. He'd be visiting Luis that evening, and he wondered if I could meet at the hospital. I had slept badly and didn't want to go.

*Okay*, I wrote. *I have some stuff for you if you have the money.*

*Cool. How much?*

*10k for the big one.*

We had not discussed price, and I thought perhaps this was pushing the limits, considering it had cost me only a grand to build.

*Alright. And the cans? You got three?*

*Yeah. $300 each.*

*Sounds good. I'll be at the hospital around eight. Top level of the parking garage. Blue Bronco. Park next to me if I get there first.*

I ordered lunch, spent twenty minutes debating taking a shower, showered, dressed, and began gathering the things I'd need to give Hector. Then I sat on the couch and watched *The Accountant*, which I had already seen five or six times.

I went to the hospital that evening around seven thirty and parked on top of the garage. Hector's Bronco wasn't there.

Luis, stoned on edibles, was, as always, pleased to see me. The two of us watched *Antiques Roadshow*, some kind of best-of retrospective. A dude with Beatles memorabilia. A Rolex worth four hundred thousand.

"What would you do with an extra four hundred grand?" Luis said.

"Beach hut in Indonesia," I said. "All my debts paid. Beautiful island girls bringing me mangos and shit. Fish, chill, walk around, get high. Then repeat the same thing every day."

Hector came in while we were discussing it.

"What would you do with an extra four hundred grand?" Luis said.

"Buy more coke," Hector said. No hesitation. "Buy more coke, sell more coke."

Luis scoffed. "This guy thinks he's El Chapo."

"Maybe he is," I said. "Could you buy a ranch for your grandfather with four hundred thousand?"

"Not any ranch I'd want," Hector said.

I went to the bathroom to let them visit while I did a line, but when I came out Hector was still on his feet.

"Can't stay long," he said. "Talk business real quick?"

"Sure. I guess I'll see you later," I said to Luis.

"You're not coming back?"

"Nah, I should head home."

Hector and I went down the elevator together and crossed the street to the parking garage.

"How's it been going?" I said.

"Not so good. Actually, it's my grandfather. He's got this Covid shit. He's already got bad lungs. Sounds like he'll probably end up in the hospital."

"Sorry to hear."

"Yeah. Maybe it'll pass."

On the roof of the garage, Hector's Bronco was parked next to the Tahoe. I opened the backseat and handed him the gun wrapped in a moving blanket. He placed it in his backseat, along with the silencers, which I'd put in a gallon Ziploc bag wrapped in paper towels.

He handed me an envelope. "This is the $11,200 you asked for, but ten for the guns is high. I know you gotta make a profit, but can you do $7,000 apiece from here on out?"

"I can do $8,000."

"Alright. I'll test this and let you know how it goes."

"I've tested it. It's good."

"Cool. I'll let you know."

In the Tahoe, I put the envelope between my legs, tucked against my nuts. It was nowhere close to the most cash I'd had on me at one time, but it brought back a feeling I hadn't had in a while. For maybe ten years, I'd dreamed of a sideline selling full-auto ARs, and now I'd done it. I'd done it only once, but once proved it could be done. If it wasn't with Hector, I could find someone else. I always figured things out. They couldn't keep a bad man down.

The moment I reached my driveway, Tonya texted. She wanted to swing by. Given the tension from the day before, I was surprised, but this is how it always was with women. You had no fucking idea what was going on. I decided to ignore the text a while. I did a line, counted the money on the kitchen counter, then stacked it up and put it in the safe in the bedroom closet with the rest of my cash. Eleven grand was far from life changing money, but it was good to see the stacks growing for the first time in a while.

Finally, I texted Tonya. She showed up twenty minutes later dressed like a temp agency secretary in a cheap black skirt, black pantyhose, and a white button-down shirt. She poured herself a glass of wine from the bottle I was working on, took a long drink, then looked at me and sighed.

"Hard day at work?" I said.

"Don't give me shit about the outfit. The guy bought it for me and had me wear it to the date."

"Secretary fantasy?"

"I wish. Weirdo wanted me to act like his sister. I've been pretending to do his taxes for like two hours. Then it went into this whole foot scenario. You don't want to know."

"Are you kidding? I live for this kind of shit."

"Apparently, his sister promised to help with his delinquent taxes. She's coming over next week, and he was pumping me for advice about how to come on to her. The guy is completely nuts. The first time I met him, it was totally normal, and every time since then has gotten worse and worse. It's always like this with guys. First, you think, 'Great, a nice regular.' Then he keeps piling on more and more weird shit, and you realize he's just another psycho."

"What did you recommend?"

"About the sister? I was like, 'Tell her how you feel if you need to, I guess, but don't like grab her or anything unless you get consent first.'"

"Aw, man. You should have told him to kiss her right when she walks in the door."

"Oh, right. I mean, as it is, I'm imagining this lady showing up to help her idiot brother, and he rapes her or something. You mind if I get in the shower?"

"You don't have to ask."

I watched TV until I heard the shower door close, did a bump off the kitchen counter, then went to the bathroom. Tonya's clothes were in a pile on the floor, a lacy pink thong and bra on top. I watched her in the shower with her eyes closed and the water pouring over her small, high breasts and her ass, firm and round. My third wife's ass had looked like a pair of deflated tires.

"You going to wear these to your next job interview?" I said.

She jumped. "Jesus! I didn't know you were there. No, I'm never wearing those again. I'm going to burn those fucking things."

"Melt them, you mean."

"Right. Polyester everything. What a cheapskate."

"Well, you're not the only one who's been dealing with unsavory characters," I said.

"Oh yeah? Gun news?"

My phone pinged.

"It's Hector. He's asking when he can get more. I guess that means he was happy with the gun I sold him today."

"Okay, don't reply," Tonya said. "You'll look eager. Get back to him tomorrow."

Together we talked through next steps. If I was going to begin producing any volume of guns, there was still a lot to figure out.

Before she left that night, she asked what I was doing for Thanksgiving.

"You're like the tenth person to ask me. You know I don't think that far in advance."

"It's on Thursday."

"This Thursday?"

"Yeah, dummy. Today is Saturday, and Thanksgiving is on Thursday."

I guess that explained why my mother had been raising the topic so relentlessly.

# 23

IT TURNED OUT Ruck was going to be in town for Thanksgiving, and he wanted to see if we could grab a beer. I hated to miss him, but there was no escaping the death march to my folks' place.

"But, hey, as long as you're in town, you think you could do me a solid and check out a junkyard for me? It's out of the way, but I'll make it worth your while."

"A junkyard? Jesus, Wince, what are you getting into?"

I'd texted Hector and told him I could get him two more ARs before leaving town for Thanksgiving. I finished converting them late Tuesday afternoon. The whole thing should have taken an hour or two, but between coke breaks, naps, smoking joints, watching movies, and taking calls from my mother, I barely pulled it off in two days.

When it came time to test the guns, I immediately dispensed with the notion of going back to Garland's ranch and dealing with Nathan and his girlfriend. In the absence of a better plan, I decided to do something stupid. I'd go to the same place Garland and I had tested my first ever machine gun. I knew it was a bad move, but I had a record of getting away with this kind of thing, and I figured if I got in and out in a hurry I would be gone before the cops showed up.

I drove the Tahoe through the creek and wove my way between some trees until I was pretty well screened from the road. With the engine running, I got out and stood listening for voices in the area, hearing only a distant leaf blower. Back when I was with Garland, we didn't give it a second thought. It seemed reckless, maybe, but nothing more. This time I wasn't so sure. I hadn't even thought to bring decoy guns like last time.

Fuck it.

I reached down and smeared some mud on my license plates, put on some ear protection, and took out the first of the new ARs.

I pulled the trigger. After five pounding seconds, the magazine was cashed. I tossed it in the backseat, grabbed the other gun, ran through a second magazine, tossed it in the backseat, threw a blanket over both of them and hauled ass.

I was at the spot for less than five minutes, but when I hit the first stoplight, two lit up police cars screamed onto the road. One of the cops hit his brakes long enough to look my way, but in the next instant he hit the gas and sailed past.

"Well, that was fucking stupid," I said.

My luck had held, but I wondered if it was thinning out, narrowing little by little.

A few hours later, Hector pulled in beside me at the top of the hospital parking garage. We got out and shook hands. For the first time since we'd met, Hector looked bad. His face was unshaven, and he had bags under his eyes. He still wore a button-up shirt, but the tails were untucked, and his starched jeans looked like they'd gotten more than a day's wear.

"You doing alright?" I said.

"Not great. I drove back and forth from the Valley overnight."

"Last night? You haven't slept?"

"Tried to get into the hospital to see my grandfather, but the fucking Covid rules don't allow it."

"How's he doing?"

"Not so good."

"Sorry to hear it."

"Yeah, you got three for me today?"

"Three? We said two."

"No, man. We said three."

"Two is what I got."

Hector looked at me like I was mistaken.

"Check your texts if you want," I said.

He took out his phone, looked at our exchange and said, "Yeah, two. You got another one you can sell me?"

"No, not at the moment."

"Alright, man. I need to take $8,000 out of the envelope. Unless you want to owe me one."

"Thanks, I'd rather you pay me for what you're getting."

Hector went to the driver's seat and messed with the cash while I loaded the ARs into the back of his Bronco. When he got out, he handed me the envelope.

"Sixteen thousand even," he said. "When can you get me more?"

"I think I could get you another three before Christmas."

"That's it? Three before Christmas? You selling to other people or something?"

That gave me pause. I'd never told Hector that every gun I made was for him, and I sure as hell didn't like him thinking of me as his personal firearm factory. I looked at him and cocked my head like a milder version of "what the fuck did you just ask me?"

"Alright, man, that's fine," he said. "Three before Christmas. But we gotta talk about the future soon."

It was exactly the kind of shit I hated. You sold drugs to someone, they thought you were supposed to be available on command. You started dating a chick, she thought you were supposed to answer the phone anytime she called. Bosses, wives, my folks. The last thing I needed was a new obligation. The part that bugged me most was that I was pretty fucking sure I needed Hector more than he needed me.

# VARIOUS HEROICS

I remember all the heroic shit I've done. I wasn't a bad guy all the time. Every time I've passed a traffic accident where the cops weren't already on the scene, I've stopped.

Once, I pulled a little girl out of a crash on the highway. The car was upside down, and her dad was dead beside her. The little girl—her name was Heather—spent about a month in the hospital after the accident. For years, her mom used to send me cards at Christmas.

Once, I was at a house buying coke when this dude OD'd. Everyone was standing there looking at the guy. No one was doing anything. I dragged him outside and across the street and gave him chest compressions until an ambulance showed up.

Once, when I was driving from California back to Texas to see my folks, I passed this barn on fire. It was a Sunday, and there weren't any cars at the house beside it. I got out and called the fire department. I didn't even know how to tell them where I was. I had to look at a map spread out on the hood of my car. Finally, I told them to go to such and such a highway and look for smoke. There was plenty of it. After I hung up, I stood there waiting when I noticed sounds coming from the other side of the barn. Horses. The fire hadn't reached them yet. I opened the stable doors, and they ran out past me. Now that felt fucking heroic.

Once, I was visiting my buddy, sleeping on his couch. He had this wild place he was renovating. Somehow, he'd found an old spiral staircase and lashed it to the side of his house with straps. It wasn't something meant to be climbed on. He was just seeing how it looked before he installed it. That night I was awakened by a huge crash. I looked outside, and there were these three college kids out there standing around a dude lying on the sidewalk beneath the staircase. They were drunk, walking home from a bar when the guy saw the staircase and, goofing around, went to climb it. He was almost to the top before it came unstrapped from the house and sent him crashing to the ground. I ran downstairs, shirtless in jeans, and held the guy in my arms while he died. His friends stood there crying. It was on the news and everything, a real sad story.

I've got a million of them—times I did something when no one else was doing anything.

# 24

I ALWAYS FIGURED Sam and I would end up having holiday traditions like my family did when I was a kid. I liked the idea of taking Christmas boxes down from the attic the day after Thanksgiving. Having a big day of decorating and playing holiday music. Rolling my eyes at my wife stringing popcorn and making me drive around looking at Christmas lights. Special recipes you only made at certain times of year. But, of course, things didn't turn out that way.

Sam had spent some holidays with friends and boyfriends over the years, and this year she was doing Thanksgiving with her boyfriend's family, of course. Who knew what Venezuelans thought of Thanksgiving? They were lucky to have Sam there to explain the Pilgrims. I halfway hoped she'd invite me to join them—a big family meet and greet before the engagement. But she didn't bring it up, so I knew I'd end up going to my parents' house in Arkansas.

Wednesday morning, my alarm went off at nine. I was out of bed at eleven. By two, I'd begun the eight-hour drive. I debated taking some coke with me, but at the last minute decided to do the trip sober. Not just because my parents are notorious snoops, but because I thought it might not be a bad idea to slow down for a while.

The drive felt good, actually. I spent the whole time listening to Judas Priest and Black Sabbath. I'd been thinking a lot lately about the underappreciated Dio albums, and I listened to them all. I decided when I got back, I'd play them for Tonya and see if she understood why they were so incredible. I liked being in the middle of nowhere listening to music. The only thing that spoiled it was my mother texting and calling every half hour. By the time I got to their place, I wanted to throw the phone out the fucking window. It was late, and I was so tired of talking to her that when I saw her standing in the driveway, I felt it wouldn't have been unreasonable to turn around and drive home.

It was my mother's house. That's how I thought of it, though my father lived there too, of course. My mother was the one who had made all the money. She was the one who had picked out all the furniture. She was the one who directed a squad of house-cleaners every week, watching as they lifted and dusted each item on every shelf. And still, without fail, my mother found something to bitch about when they were gone. My father picked out his boat, his pickup, and his beer. The rest he left to her, which was the wisest strategy, considering.

Her house was a different world than the one where I grew up. Not that I was raised in a mud hut exactly, but it was nothing like this. That my mother had risen to a life on this sprawling property in an exclusive lakeside community was only surprising if you didn't know her. The story of a pregnant fifteen-year-old girl from a shitty neighborhood in a poor Texas town ordinarily lends itself to a certain range of outcomes.

Maybe a woman in that position, if she was a real striver, would put herself through nursing school or retire from a long career as a tax preparer. But if you knew my mother, you wouldn't marvel at how far she had come. You would wonder at the limitlessness of what she would have achieved under any other circumstance. My whole life she rose to the top wherever she worked, but it wasn't until she started selling life insurance that she really found her groove. All that networking, all that pushing, all that concern mixed with pragmatism. My mother saw all of that and recognized how she could build an engine that would power her whole life forever.

My mother's house is probably sixty-five hundred square feet, and every inch looks like it's straight out of the Ethan Allen catalog. White couches with little tan throw blankets, ivory rugs, matching blue and white lamps, beige walls covered in the kind of art you forget the second you see it. Bookshelves staged with off-white vases and neutral-colored books no one ever opens. It's like a model home.

Though they'd been married for as many years as I had been alive, my parents were still in love. It was incredible to me that they had been so lucky, and I envied them. All I ever wanted was to find something that permanent and lasting, where the two of you stick it out through being broke and being rich and whatever other problems you run into.

Every night, my dad sat working his way through a six-pack of Budweiser in front of the TV while my mother ran around doing all the little tasks she'd taken up after retirement.

My mother talked as I unloaded the Tahoe, and she talked as I dumped my stuff in my room, and she talked as I greeted my father. She told me about the various things her neighbors were doing for Thanksgiving, her recent scores on the golf course, and about a singer-songwriter she and my father had seen at a café in town. It was always this way with her, a ceaseless stream of information. I never knew how to parse which parts were meant to be important, and she never gave any hints.

"God, I'm hungry," I said at last. "You have anything to eat?"

They were magic words. She was off to the kitchen, piling plates with whatever she had stored up in preparation for my visit.

My father had stayed up late to see me. He gave me a little smile as I sank into the sofa beside him.

"Want a beer?" he said.

"Nah, I'm okay."

"She's been excited for your visit."

"I wouldn't have guessed."

"Drive good?"

"Not bad. How about you? How's your day?"

"You're looking at it. You want a beer?"

"No, thanks. I'm okay. Hey, when Mom's not around later, I could use your help with something she wouldn't want you involved with."

"That sounds exciting," he said without a trace of excitement. "I'll make sure and show you the boat tomorrow after lunch. I'm going to bed. If you want a beer, they're in the fridge."

There was no such thing as sleeping late at my mother's house. God knows, I'd tried. For years, during every visit I'd fake a mild illness, hoping she'd let me sleep in. She was always vigilant about health, which was probably why my father was still alive after multiple bouts of cancer and a heart condition. After I slept until noon for a third holiday in a row, my mother became convinced I had leukemia. It was easier getting up the first or second time she popped her head in the door. Thanksgiving morning I was out of bed by nine, zombie walking across the plush, white carpet.

I went to the kitchen and found her already racing from one oven to another. She'd hired a woman to help with lunch. It was a different woman than the year before, but as always, my mother seemed to know enough about her to inquire about the health of her grandchildren.

I took a Coke from the fridge—my mother always stocked them when she knew I was coming—and went to the living room. My father was there looking at his iPad having his third or fourth cup of coffee.

"Sleep okay?"

"Yeah, you?"

"Yeah."

My phone pinged. Luis texted a gif of a cartoon turkey dancing with a pilgrim. When I texted him back, he told me Hector's grandfather had died.

*Aw, man, sorry to hear that*, I said.

*It happens*, Luis said. *Hector said he was pretty impressed with the products you sold him.*

I stopped replying. Even by the standards of a normal civilian, Luis was too loose-lipped over text. By the standards of a drug dealer, he was completely off the rails.

Bryce texted around two, by which point my father and I were on the couch watching an episode of *Pawn Stars* and waiting for Thanksgiving lunch to be served.

*Where are you?* the text said.

*Who's this?*

*You go to Arkansas?*

*What do you want?*

*If you're in town, come to the boathouse later.*

Bryce had built a ridiculously extravagant house on the lake. He and his skinny wife kept adding shit like heated floors and walls of glass, and he kept firing contractors, so it took almost three years to build. I'd still never even seen the whole place. His boathouse was set up as a combination yacht club and tree house, dark paneled walls and a walk-in humidor paired with a full-scale pinball machine and an autographed *Scarface* poster.

*Family is out of town,* he wrote. *If you're too fucked up to drive, I'll send an Uber.*

I didn't reply, but it felt good knowing Bryce was alone on Thanksgiving. It wasn't even three in the afternoon, and he was already bombed. I was halfway jealous of the bastard.

Thanksgiving has always been my least favorite meal of the year, this weird mix of the same gross, mushy foods. Turkey, green bean casserole, dressing, yams. The spiral cut ham was alright, but I could have lived my whole life without eating the rest of that shit. Like always, I choked down what I could and tried not to complain. My father repeatedly told my mother how good everything tasted, and she replied by saying that the turkey was too dry, the green beans too soft, and the pecan pie too sweet.

"I read something in the paper this week about how important it is to check in on each other during all this Covid business," she said. "A lot of people are having a hard time, it seems like."

"Yeah," I said. "I've heard that."

"The thing I read said when you ask people how they are and they say they're fine, because that's what people always say, you should ask it again but this time you say, 'How are you *really?*'"

"Huh," I said. "Interesting."

"So, I was thinking maybe I don't ask you that question often enough. I call and get started in on everything that's going on with me, but I wanted to ask you while we're all together. Honey, how are you *really?*"

My first reaction was to scoff a little, but the question did give me pause. I sat there a moment blinking at the orange candles nestled among the plastic autumn leaves of the centerpiece, feeling myself getting a little choked up out of nowhere. "I guess I'm not doing so great, Mom, if you want to know the truth."

She nodded, wiped her mouth with her napkin, and put it back in her lap. She reached out and put her hand on top of mine, her cool rings against my knuckles. "Not drugs again?"

"No, nothing like that. You know me. I've just been off my game since I left Keystone. And I miss Sam. I mean, I'm glad she's grown up and happy with the Venezuelan and all, but I miss when she was a kid. I don't know how to explain it. It's like I was rising and rising and then I flew into a wall."

"Honey, you have plenty to offer. You're so smart. You can do anything."

"That's right," my father said, slapping my knee under the table.

"Maybe so," I said. "But starting over at fifty-seven?"

"We know how it is," my father said.

It was true. They'd lost it all more than once and had made it all back and then some.

"Gotta stick it out," he said.

"Ever think of going back to Keystone?" Mom said. "They're in a better position now, you said."

"Yeah, they are. But I can't go back to work with Bryce. Not after all this shit."

"Wonder if it would help if you had a girlfriend," she said. "I can't remember you ever going this long without one."

"That's the last thing I need right now. Considering the chicks I've been picking."

My mother, who had hated practically every woman I had ever been with, tilted her head in acknowledgement of my monumentally poor taste.

"Don't you have any projects?" she said. "Your whole life you've had projects. Motorcycles and boats and cars, hunting.

All those built-in shelves at the old house. I never hear you talk about things like that anymore."

"To be completely honest, Mom, right now I'm having trouble even getting out of bed in the morning. I'm sleeping half the day. Then I go back and take another nap. I'm not going out, not seeing friends. I'm not in a good place."

It was the frankest I'd been with my parents since the rehab years, and I braced for the inevitable bootstraps speech.

"Oh, no, honey. I'm sorry to hear that. I'm glad I asked then," she said. "And I'm sorry I didn't say something sooner when I had a feeling something was wrong."

"I don't need you worrying about me. I'll get back on track."

"Well, you always do, honey. You always land on your feet. What can we do to help?"

"Maybe don't give me any shit if I'm not at my best," I said. "Or if I don't feel like talking on the phone sometimes."

"Everyone goes through hard times," my father said. "The first time we lost our asses and we were back living with your mom's parents, I thought the whole world was over."

"That's right," she said. "But we got past it. Now look at us. I wonder if you're depressed."

I almost laughed. "I'm definitely depressed. I've been on Prozac for almost two years."

"Doesn't that stuff mess with your pecker?" my father said.

"Well, it ain't the best thing that ever happened to it."

"Honey," my mother said, "I don't think you should do anything that affects your erections. That's enough to make anyone depressed."

When we got up and carried our plates to the kitchen, the woman my mother hired stood wiping down the countertops pretending she hadn't overheard us talking about how I was a medicated, impotent loser. We piled our dishes in the sink, and my mother hugged me and looked up at me.

"Did I ever tell you, Irma, how proud I am of my son, Winston?"

"Yes, ma'am. You told me he's a real good boy."

"That's right," my mother said. "And handsome, too, isn't he?"

"Very handsome."

"I'm going to show Winston the boat," my father said.

"I'll help Irma tidy up," my mother said. "I'm sure she wants to get home to her family."

Outside, I lit a joint and offered a puff to my dad, as I had done many times since I started openly smoking pot around him in my early thirties. He refused the joint, as he always did and popped the top on a Budweiser he'd grabbed on the way out of the kitchen.

"So, what do you want my help with?" he said.

"I need you to order some gun parts for me. AR-15 parts. I'm trying to put a bunch of guns together."

"Sure, we can order them before you leave. You having money problems?"

"No. Paying for them isn't the problem. I just need you to order them, have them sent here, and then mail them to me. I'll pay you back. I know it's a pain in the ass, but I've ordered a few lately, and I don't want my name on any more lists than it's already on. You know how the fucking ATF is."

"Mmm," my father said.

"You haven't bought any guns lately, have you?"

"Maybe at a garage sale last year, but that's it. Nothing on the record."

They didn't make people like my dad anymore, who didn't pry or interrogate or press.

Later that night, my father and I went into his study to buy the parts I needed. Right away, there was a problem. I had been so fixated on staying under the radar, it hadn't occurred to me that, due to supply chain issues, parts actually were hard to come by. It wasn't just something people were saying. There were a handful of sites I trusted, but even those were so low on stock they were limiting the quantities any one customer could order. We ended up getting everything I needed, but it required ordering from five different websites, plus two-day shipping, which cost me hundreds more. The whole thing was such a pain in the ass my father went to bed before we finished, leaving me with his credit card.

My mother knocked on my bedroom door at eight o'clock Friday morning. I rolled over and looked at my phone and found a message Tonya had sent at two a.m.

*Hey, you didn't, by any chance, come home tonight, did you?*

I spent the day helping the old man with the boat and my mother with the television set. Every time I did something any twelve-year-old could have handled, my mother praised me to the heavens. What a smart, good boy I was. I had to admit it did feel good.

For a minute, I even entertained the thought of moving in with them. That would solve a lot of problems. Plus, I could help take care of Dad, whose mental decline was way more pronounced than it had been even just a few months earlier. Twice, he asked if I needed help ordering the gun parts we'd ordered the night before. It wasn't such a loser move if your folks really needed you. I hadn't done coke in nearly forty-eight hours.

I wanted to show them *The Accountant*, so we watched it together, and we all agreed it was an excellent movie. After my parents went to bed, I smoked a joint on the patio outside my room and looked up at the stars. Maybe it wouldn't be such a terrible life.

The next morning—a Saturday for God's sake—my mother started banging on my door at seven thirty sharp. "Come get your breakfast. Your eggs are getting cold."

I was on the road by ten.

I figured if I got home early enough, I might catch Ruck and hear what he'd found at the junkyard.

# 25

THE MOMENT I got back, I texted Ruck.

*Got home a day earlier than planned. You still in town?*

*Yep. Wife's family is crazy as ever. Grab a beer?*

The place we met called itself a tavern, but it was a dive in a strip center. I got us a table and sat nursing a Don Julio añejo, neat. Ruck came in looking even fatter than I expected. He spotted me, smiled and nodded, then went to the bar to get a drink.

I watched him waiting to be served. It wasn't exactly a young, hip place, but he was still the oldest guy in the bar, with his grey beard and his Lee jeans. He wasn't wearing his biker vest like I'd imagined. He had on a plaid flannel, tucked in over what they used to call a dunlap belly, because it "done lapped" over your belt.

When he got to the table with his bottle of Miller Lite, he stood by my chair like he wanted me to stand up and give him a hug, so I did.

"What happened to skinny Wince?" he said.

"You're one to talk."

When we sat down, he took his phone from his pocket and opened his pictures. He swiped past Thanksgiving photos until he got to the junkyard shots and slid the phone across the table.

"How was it?" I said.

"Not bad. I took my wife's nephew. He likes cars. He didn't even know there were actual junkyards that let you pull your own parts."

"That's kids nowadays. You talk to anyone who works there?"

"Some dude at the front office. I found a bumper I wanted, and the guy said it'd be a hundred and twenty."

"Mexican guy?"

"Yeah. Probably twenty-five. Mustache."

The pictures weren't very good, but they were helpful enough to make the favor worthwhile.

"So, what are they selling out there?" Ruck said.

"You saw the place. They're selling junk. For too much, it sounds like. I don't see any pictures of the mobile home in back."

Ruck took the phone and swiped. "There. Front and back. So, you've been out there yourself?"

"Once at night. As a driver. Helping out a guy I know. By the way, you know anybody who'd want full-auto ARs? I got some I'm trying to sell."

Ruck took a long drink and looked at me.

"What?" I said.

"I'm wondering what it takes to turn a banker into a one-man crime spree."

"Not that much, it turns out."

"When are you thinking of going out there?"

"I don't know, man. Maybe never. It's just a thought."

"Not like you to do too much thinking," Ruck said.

That got a smile out of me. "Fuck you, old man."

Ruck and I hung around drinking for a couple of hours. I was feeling pretty good when I left. Ruck hugged me in the parking lot and said, "I love you, man."

On the way home, I texted Tonya. She responded hours later and told me she was busy. I didn't press or ask why she had been looking for a place to be on Thanksgiving night. I needed time to recover from the visit with my folks, anyway. Which for me meant getting as much sleep as I could. The next day, I lied to Luis that I didn't know when I'd be getting back. I slept and ordered takeout, watched movies, and hardly did any coke.

Right when I was starting to feel the slightest twinge of loneliness, Tonya texted. It was Wednesday afternoon. I hadn't seen her in more than a week.

*What are you doing?*

*In bed.*

*Want a napping partner?*

*I haven't showered.*

*Then shower. Or don't. I'll be there in 30.*

A more complete disclosure on my part would have been I haven't showered since Thursday. I hauled myself out of bed, unlocked the front door, did my last two bumps off the kitchen counter, and turned on the shower. By the time Tonya arrived, I had almost fallen back to sleep with the sound machine set to waves. When she came in, I didn't roll over or speak to greet her.

I listened to her undressing and felt her get into bed with me. She put her arms around me and inhaled the coconut shampoo in what was left of my hair.

"You bathed," she said.

"Don't get a big head."

She laughed.

"I have something to confess," I said.

"If it's something that's going to hurt my feelings, I'd rather not hear."

"I stole something from my father this weekend. Three Cialis tablets."

"Oh, honey. I'm really not in the mood. I've had the worst week. Can we have a normal night? Like, sleep for a while, and then you can complain about everything that's happened since the last time I saw you?"

"Complain? You think I complain?"

"You complain constantly."

"If you met my mother, you'd really see a complainer."

"Where do you think you learned it? Anyway, I don't mind when you complain."

We napped, and for once I was the first to get up. I took hits off the bong on the kitchen counter, ordered Chinese, and watched an episode of *Bernie Mac* in the living room. When the food arrived, I woke Tonya, and she sat at the table looking bleary, eating, and not talking much.

"Thanks for the fried rice," she said. "I like this better than the other place you order from. No peas."

"Yeah, I remembered. You doing okay? Rough holiday?"

"I don't feel much like talking about it, to be honest."

"Cool by me."

So, as usual, I did the talking. I told her about the nights I'd spent with my family, and though I could hear how the things I said might sound like complaints, she laughed as I talked, so I kept on. I told her about how my folks surprised me when I came clean about how depressed I'd been.

"They were actually pretty fucking cool, for once."

"Sometimes it pays to have a little faith in people."

"Maybe," I said. "You feel like watching a movie or something?"

"Thanks, but I should go. We'll try the Cialis another day. But first I'm going to do the dishes. The pile in your sink is grossing me out."

It was something she did more and more, picking up around my place. Sometimes it bothered me, but it was always nicer when she left and the dishes were put away or the stacks of junk mail had been sorted. When she was finished, even the countertops were wiped clean.

"I'd like to tackle the fridge," she said. "But I really don't have it in me today."

I walked her to the door, and, as she always did, she kissed me on the lips. I told her to drive safe, and off she went to who knows where.

## 26

I KNEW HECTOR was out of town the week after Thanksgiving, attending to the details of his grandfather's funeral, which gave me time to complete the next three ARs I'd promised him. And he'd also texted to ask if I could get him five more silencers. I told him it wouldn't be a problem and ordered ten solvent traps for overnight delivery.

I was out of coke. Luis had been released from the hospital the day after Thanksgiving, and I still hadn't gone to see him. I called on the way over to ask if he wanted me to grab him a burger, but he told me he was making breakfast for supper, and that he'd make me some, too.

When I got to his house, he was in the kitchen, hobbling around in a plaster cast, having gathered his supplies: butter, eggs, a slab of bacon, grated cheese, salsa, and tortillas.

"Hey, vato!" he said. "Man, you called at the perfect time. I'm about to make us a breakfast feast. Like I used to tell the boys, breakfast is even better at night."

Now that he was home, I had the impression Luis's weed and coke intake was that of a guy making up for lost time, and it wasn't hard to imagine some version of the same thing happening again. Not that I judged.

"Being in the hospital that long is weird, man," he said when we were seated in our usual spots at his breakfast bar. "It's like your life is on pause at first because it's just a day or two, but then it's like your life starts all over again. It's kinda spooky."

"Yeah, I hear you, man."

"I wish we could move to Indonesia, like you said. Huts on the beach. Beautiful girls bringing us mangos and shit."

"Yeah. Except I've been researching the drug laws there. Pretty fucking brutal. They catch you with enough coke, you don't end up in fucking Indonesian prison, they cut your head off. Even for weed it's like twenty years or something."

"Fuck that," he said. "Well, maybe we'll think of another place. Somewhere with the beach and the girls and all that. Or maybe we go and don't get caught."

Don't get caught. Everybody thought that way, like they could decide to go unnoticed for their criminal activity. Luis had already spent five years locked up in Huntsville for selling coke. You think he'd know better.

We had a killer breakfast supper, but when I left Luis's with two grams of blow in my breast pocket, I was feeling pretty low. I figured I knew how he felt, like maybe the whole medical episode had seemed like it was going to change his life, and now it hadn't. The way I saw it, Luis had been permanently installed in this existence of his, and it was always going to be the same until renal failure or a heart attack took him out of the game. What bugged me most was the question of whether the two of us were stuck in the same loop together. There were no guarantees, in fact, that he wouldn't outlive me. You never knew how close you were to one cigarette too many, one dessert too many, one snort too many. I tried not thinking about it.

That evening, Sam called. Sometimes hearing her voice was like a buoy that lifted me to the surface. We talked for ten or fifteen minutes about what was happening at her job at the YMCA, where she was teaching kids how to swim. She told me about her classes, and what was new with her soon-to-be fiancé, whose name, I had to keep reminding myself, was Daniel. Soon, I'd have to learn to stop thinking of him as the Venezuelan.

The issue of the ring had finally been resolved over Thanksgiving, and my mother had given her blessing for Sam to do as she pleased with the stones from her mother's ring, on the condition that she returned the original band and setting. Sam had a feeling Daniel would propose on Christmas, she told me. Which I took to mean they had planned it. I hoped I would never hear about the fucking ring again in my life, but I knew the subject almost didn't matter. There would always be another set of problems, and another and another and another until I died.

The new AR-15s were no exception. Once again, I had to figure out how to test them. I saw on Facebook that Garland and Candace were back from Mexico. I could have gone out to their ranch for a big reunion hangout. I knew Candace would cook me up an amazing dinner. In the abstract, it sounded good, like something I'd like to recount later. It would have been cool to go to Luis's house and tell him how I'd gone out to my old buddy's ranch where we'd caught up and smoked some dope and shot some guns. And how his wife, whose ass I used to grab, told me about all the pretty girls she could set me up with.

Recounting it seemed fun but living it didn't. I didn't want to go out there and talk about politics and Covid and hear about their time in Mexico. I figured Nathan would be there again with his pudgy girlfriend. I couldn't understand why my feelings towards them all had changed. Nathan had once been like a son to me. In fact, I was pretty sure I was his godfather. It bugged me that I didn't want to go to the ranch, but not enough to go.

I didn't consider going to that spot over the creek near my house again. The cops had appeared pretty much instantly last time, and if they had shown even the slightest bit of interest in me I would have been fucked. A decade ago, you could fire a gun or two without everyone having an absolute meltdown. Street signs still had bullet holes in them. Now, it felt impossible.

I couldn't take illegal, fully automatic AR-15s to a firing range if I wanted to. Not that I would want to. It always surprised people how much I disliked shooting ranges. Those places were a drag even if you were testing a legal weapon. A bunch of wannabe tough guys and tourists whose visits to Texas wouldn't be complete unless they fired a real live gun while they were in town. And it was like a law that the guys who worked there were always assholes with all these dumb rules about how everything had to be.

Shooting outdoors was the only way to go, as far as I was concerned. It was like pissing outside. It felt more natural and fun.[1] Finally, I loaded the three AR-15s into the Tahoe, did lines off the kitchen counter, and started driving, figuring I'd find a place. It was the middle of the day, and everyone was at work except me. I could head in any direction, and I'd find some solitude eventually. I liked that about Texas, that you could drive forever. I mean, it could be a curse when you were trying to get the hell out of there, but most of the time it was nice. So, I drove, heading in the direction where I was least likely to run into civilization.

After forty-five minutes, I was completely lost. I could have looked at the map on my phone, but what difference did it make? Every time I saw a road that seemed to lead nowhere, I took it. Every time I saw a house, I took the next turn, trying to get as far away from people as I could. It had been a long time since I'd done that. It was a trip being out there roaming around with no destination, following the road to wherever it led. You could forget there was so much nothingness so close. It didn't even take an hour to get there if you treated civilization like an inverse compass, heading away from its markers.

As I knew it would at some point, a dirt road I took finally dead ended into a heap of trash. If you take enough turns on any back road, you end up at a burnt-out old couch, a pile of tires,

---

[1] The problem with Texas was that there was no Bureau of Land Management like you find in the Western states. Or, anyhow, the only tract of it I ever heard about was up near Amarillo. Other places, you could go to all sorts of public land to hike and camp and shoot. But in Texas, someone owned every piece of land you stepped onto. If you weren't in a city park or something, you were on some guy's property. And even if that guy gave you the go ahead to fire whatever the hell you wanted, in Texas you had to be on at least fifty acres to shoot a gun in city limits. And you had to be more than 300 feet from the neighboring residence. And you had to be "reasonably confident" your bullet wouldn't end up on someone else's land.

and scattered black bags spilling out all the shit nobody wants anymore. I didn't know which county I was in, but I figured that unless I was having an extremely unlucky day, there was no way the cops would make it out there by the time I was finished.

I turned the Tahoe around, took out the first AR, pulled the charging handle, and fired into the couch. It worked perfectly, and the couch was shredded. I tested the second into a stack of tires. I was starting to wonder why I even bothered with this testing routine at all.

Then I tested the third and got a reminder. I pulled the trigger, and nothing happened. I checked the selector to confirm it was switched to AUTO, and found it was. I switched it to SEMI and fired. It worked. One bullet came out and punched through a black trash bag. I switched it back to AUTO and pulled the trigger again. Nothing. Shit.

# 27

AT HOME, I found the problem with the third AR. I had milled the shelf to the wrong depth by a cunt hair. It took five minutes to adjust it. It was exactly the kind of thing that made me nuts. Not because it was a stupid mistake that meant I'd have to keep testing the fucking things, but I kept thinking of all those dumb politicians going on about how any twelve-year-old could make a machine gun in their garage in fifteen minutes. Good fucking luck.

Hector texted the next day, and I halfway debated delaying the sale to test the gun again, but I wasn't going to go on another endless drive, and I figured it was fine after the work I'd done.

I met Hector at Luis's house that night to sell him the three ARs and five silencers. The two of us excused ourselves to Luis's garage to talk. It was a musty, junked-up one-and-a-half-car with crumbling sheetrock and boxes of shit everywhere.

"You tested these?" Hector said. "I can sell them without taking them out myself?"

"Yeah, they're tested. Good to go for anyone."

"Cool. Hey, I've got a pick-up at the junkyard later this week. You're welcome to join if you feel like driving again. I could kick you a half ounce."

I pretended to consider his proposition. "Nah, man. I got a lot going on right now. Maybe another time."

"Alright. How many more can you get me?"

"How many do you want?"

"Can you do ten by next week?"

"Ten?"

Before that moment, I still believed it was possible Hector was moving the guns to his homies, guys blowing their cash on shit they could show off to their friends but would probably never use. But now I understood he was moving them to his cartel connection. I guess I should have assumed, at the rates I was selling them, they were going to people who cared more about firepower than they did about money. Or else he was passing them along at his cost to make inroads with some heavy hitters.

"I'd need more than a week for ten," I said. "Two weeks, at least. By the way, I've been meaning to ask, if these guys need to get armed so quickly, why don't they use lightning links?"[1]

---

[1] There's a less expensive and complex way of achieving almost the same effect as a full-auto conversion, which is something known as a lightning link or swift link. It's essentially a piece of metal you insert into the lower receiver that allows the gun to operate as full-auto. But there are trade-offs. One, you don't have a selector switch. The gun operates in one mode: full-auto. Basically, it instantly becomes a bullet firehose. The other problem is that most of them are made of thin sheet metal and probably wouldn't last much more than a couple thousand rounds. For a while, one company was even selling them as keychains, but I think that got shut down pretty quick. Now people were 3D printing them. Those lasted probably only 750 to 1,000 rounds, but they worked in a pinch. It was something I'd been wondering about since I'd first had the idea to start converting ARs. Why pay a fortune for a full conversion when you basically get the job done for less than a hundred bucks and toss the evidence before you got busted?

"Same reason someone wants a Cadillac instead of a Toyota," Hector said. "Plus, you know, without a selector switch, it ain't the same. A swiftie is cool if you need to walk into a place and fucking spray, but fully converted is better for a lot of guys. They wanna be able to blast, but they also want to take aim and fire off shots without having to reload every five seconds."

"I wonder if these hombres know what they could be getting for a hundred bucks."

Hector smiled. "Appreciate you being so concerned about the customers, but, trust me, this is what they want."

# L.A. SHOOTOUT

I remember a time in L.A. My first wife and I were living with her parents in this shitty little house with walls about as thick as wax paper. Not an easy place to have a drug habit. I went down to Watts to buy some freebase. That's what I'd do then, walk down the sidewalk in a shitty part of town waiting for a dealer to approach me. Maybe it made me an idiot, but I was never scared. I ended up buying the shit from this Black dude who was probably forty but dressed like he was fifteen, in a big red basketball jersey and shorts. I needed somewhere to go that wasn't my in-laws' house, so I told him I'd pay for him to get fucked up too if he could find us a place to do it.

I followed him to a garage he had set up with a couch and a TV, and the two of us lit up. He explained that it was his buddy's place, but he was allowed to hang out there and even sleep there if he wanted. It didn't matter to me as long as I could get high.

Smoking coke is a lot different than snorting it. It was my one big weakness, and, even then, I knew I'd have to quit soon.

We were sitting there, zoning out when two guys busted in on us. Big Black dudes with pump shotguns, yelling and shit. I didn't know what was going on, but I knew it didn't have anything to do with me. I stood up, and said, "Hey, man, this is—" And one of the guys whacked me in the mouth with the butt of his gun, a Mossberg 500, I think. It busted my lip wide open. I fell back into the couch, and the two guys turned their attention to the guy who had brought me there.

As soon as they looked away, I drew an old Smith and Wesson snub-nose I used to carry. I shot both of them. At least, I think I did. I know I got one in the neck. The other I thought I got in the chest, but I was up and running before the bullet even hit him.

I had to see a plastic surgeon to fix my lip. I told my wife and her folks I'd been mugged after work. I watched the news at the hospital, wondering if I'd see something to tell me it really happened, but no one ever mentioned it.

# 28

MY PARENTS HAD Covid back in July. Sam had it during the first wave. Luis swore he had it three months before any of us knew the fucking thing existed. More than half of the people I was friends with on Facebook had posted pictures of their stupid little test showing two lines with notes like, "Through God all things are possible" or "Thanks, Obama" or "China ain't killing this motherfucker." The guy who posted that last one actually did die of it.

Somehow, I never got it. I was starting to think I had some kind of natural immunity until three days after seeing Hector, I started feeling very, very unwell.

I got this incredible headache that kept building and building. No Advil or Tylenol could touch it.

My mother called, and I could barely make sense of what she was saying. When, gradually, she grasped that my zoned-out affect was due to delirium rather than my usual disinterest, she went into full Mom mode, insisting I take a Covid test right away, threatening to drive down to take care of me. That was smelling salt enough for me to come out of my haze, at least long enough to persuade her that I'd be fine on my own. I didn't have any tests, so she overnighted a crate of them. By the time they arrived, the results were a foregone conclusion. The headache had subsided by then. I'd moved into a new phase—the feverish, snotty, coughing, wheezing, sweating, exhaustion part.

Tonya texted to see if I was available.

*I'm sick*, I texted back. *Covid. I wouldn't come if I were you.*

She came. She wore a red and white striped shirt, a pair of black jeans, and a surgical mask. She'd stopped at the Walgreens near my house to buy cans of chicken noodle, saltines, Nyquil, Advil, two boxes of Kleenex, a case of bottled water, and two bottles of Cabernet.

"Jesus Christ, you're an angel," I said.

"You're stoned."

"Who wants to be sick and sober?"

"How do you feel?"

"Like hammered dog shit."

"You look worse than that."

She warmed a bowl of soup, which I ate with a dish towel stuffed into the front of my shirt while she did the dishes, tossed out the fast food garbage on the counters and in the fridge, and took the can from the garage to the curb. When she came back, she wasn't wearing her mask.

"Hey, put your mask on. You're gonna get sick."

"What am I going to do, avoid you for two weeks? At least we'll be sick at the same time."

"Don't be stupid."

"I'm young and healthy. It won't be as bad for me. Anyway, I'm sick of worrying about it."

She got a blanket from the bedroom and sat next to me. We watched TV for a while. I fell in and out of sleep until she walked me to bed, kissed me on the lips, and let herself out.

When I saw her three days later, she was red nosed and sniffling. She rested her head on my shoulder. "You got me sick."

"I warned you."

"It's fine. I got a booster. You didn't, I'm guessing."

"I had one shot at the beginning. That's all they're getting out of me."

"Sometimes I forget how stupid you can be."

"You're the one who got it on purpose. Don't you have a roommate or something? Someone else you could get sick?"

"I have a husband, actually. I've been meaning to tell you."

The way she said it didn't sound like a joke. She never lifted her head from my shoulder.

"You really have a husband?"

"For the moment."

"What does that mean?"

"You're the guy who's been divorced three times. I could use some advice, actually."

"Advice about what?"

"You know what? Let's not talk about it tonight, okay? I've been wanting to tell you. Now you know."

It wasn't until Tonya revealed that she was married that I realized how much I'd dreamed up a completely imaginary life for her based on what shreds of information she'd given me. She'd mentioned certain restaurants and shops she frequented, and when she said she was home, it always took her right around twenty minutes to get to my place, so I had a fairly good sense of where in town she lived. I'd once dated a chick who lived in an apartment complex in that part of town, so I guess I'd sort of superimposed Tonya into that place. She'd thrown me off by mentioning a roommate in the past, and because I knew hookers often lived together, I assumed her roommate was one of the girls from that scene. She knew I thought that, because in stiffer days I'd mentioned the possibility of us having her over for a threesome sometime.

"Did you get your husband sick?"

"No, I told him one of my clients thought he had it, so he's been keeping his distance."

"Maybe we shouldn't talk about this tonight."

"That's what I just said."

Later, we went to bed and coughed side by side until finally she looked at her watch and said, "I should probably go."

"When are you coming back?"

"Why do you ask?"

I didn't know how to answer her. For a long time, I'd thought of myself as the one consistent man in her life. Now that story had changed.

# 29

I WAS WELL enough after a few more days to work on the ten guns Hector had requested. Which isn't to say I did begin working on them. Covid had knocked me back into my usual slump. I had the parts I needed, thanks to my dad. And, once I finished them, I'd have eighty thousand in cash, plus what Hector had already paid me. It would be enough to get by for another eight months if I lived frugally.[1] Continuing to get parts would be a problem, but like everything else I was sure I would figure it out if I applied myself to the question. I still wondered about the clientele. It bugged me a little. If I was becoming an arms dealer, I might eventually have to deal with cartel guys myself. Hector might want to introduce me around. Or, if he got into trouble and thought it would help, he'd give them my name himself, and I wasn't sure I wanted to deal with those guys directly. I'd seen

---

[1] I had rent, which was $3,000, then another $2,000 or $3,000 in DoorDash—meals, wine, all that shit. Then I had cable, utilities, internet, insurance, credit card payments, my car payment, and Sam's car payment. Her student loans were suspended while she was in grad school, thank God. The rest of my money went to coke and cigarettes. Maybe three grand a month in coke sounds like a lot to someone who doesn't use, but it never seemed that bad to me. When I left Keystone, I had more than enough to last three years. Now, I was close to broke.

videos of what cartel guys did to their enemies. In one, a group of men cut off a man's hands and had him try to defend himself against kids armed with machetes. Mexicans in the U.S. always had so much pride about where they came from, it kind of shocked you to see how badly they'd treat each other.

I texted Hector.

*Been sick with Covid. Bad case. Getting better, but the stuff will be late.*

*Sorry to hear it. How late?*

*Not sure. At least a week. Thanks for your patience.*

An important business rule I learned many years ago: never apologize. Instead, thank people for their patience and keep things vague.

*Keep me posted. Sooner would be better.*

Hector wasn't afraid to be pushy. I didn't like it, but it did work. I got off my ass and began building the ten guns. As always, the hardest part was getting started. Having to build ten was a blessing and a burden. It was easier to find a rhythm and hit a groove with the work, but it was still a grind. I couldn't believe this was what I'd been dreaming of, working in a one-man factory with the TV on in the background.

The next time I saw Tonya, I was in the living room surrounded by guns, naked except for my big white robe. It was two in the morning, and I had done more coke in a single night than I had in at least ten years.

"Jesus Christ," she said. "You're on the verge of your River Phoenix moment. You look red."

"Do I?"

"Go look in the mirror. You look like the fucking Kool Aid man. I was hoping to try the Cialis tonight, but I think if you put one more substance in your body, your head's going to pop off."

"You ever had a client die in the sack?"

"It happened to a friend of mine."

"With a regular?"

"No, she'd never met him before." Tonya gazed around the living room. "Look at you. You got your little workshop going, huh?"

"Pretty impressive, right?"

"Not bad at all."

She stood there a moment looking at me like she was thinking about something.

"What's up?" I said.

"It seems like a bad time to mention it, but I have a favor I need to ask, unfortunately."

"Oh?"

The second she said it, I knew. I set down the gun I was working on and looked at her. Now I was listening for a figure. How much would she hit me up for? It was such a fucking cliché I could barely stand it. Why did everything have to turn out the way you always knew it would right when you thought you'd been wrong?

I started to ask how much, but that seemed too unkind.

"Why are you looking at me that way?" she said.

"Just listening. What's up?"

"Well, now I feel awkward about it."

"Spit it out, honey."

"I need to borrow some money if you can spare it."

And there it was.

"Oh, yeah? How much?"

"Five thousand. It's the retainer for my divorce attorney."

"Five thousand seems like a lot. Your husband have a bunch of assets or something?"

"Some. Retirement accounts mostly, but I don't give a shit about those or about the house. Mainly I'm worried about a family ranch. My grandfather's. My husband helped me buy out my siblings a few years ago. It should be mine, but it became ours when that happened. He's already told me he'll fight me for it."

"Where's this ranch?"

"South Texas."

"How big is it?"

"Two sections."

"Anything on it?"

"Not much. A three-bedroom house built in the 60s. Couple of cabins no one has lived in for the past twenty years. Fences that need fixing. Listen, you don't have to answer right now. Think it over."

"Okay," I said. "I will. So, you're from west Texas, huh?"

"What makes you say that?"

"You said 'sections.' That's sort of an old school West Texas term. Never heard anyone use it to describe a ranch in South Texas. Usually, people just say how many acres they have."

"Usually, people don't tell you how many acres they have at all," she said.

"That's true. My mother always told me it was rude to ask how big someone's ranch is."

"She was right. But I don't mind you asking, of course. I tell you everything."

"Ha. You don't tell me anything."

"My grandfather was from West Texas. And he was the definition of 'old school.' I guess I use that term because that's the one he used. It's a little over a thousand acres."

"Pretty big ranch for a house and a couple of cabins."

"Yeah. I've got other plans for it. Anything else you want to know?"

"No."

"Okay. Do you mind if I crash here tonight?"

"Course not. I was thinking of going to bed myself."

The truth was I was nowhere near thinking of going to bed. I'd inhaled so much blow, I assumed I'd end up watching the sun rise off my deck. But I figured what the hell. I should sleep. I took a Xanax, four Advil PMs, and a fist full of sleepy time gummies.

Tonya brushed her teeth with a toothbrush she now kept at my house, undressed, and fell asleep about two minutes after her head hit the pillow. She was snuggling the one that was supposed to go under my legs. Maybe that's why I couldn't get comfortable.

I took my phone from the nightstand and scrolled Facebook. At around 3:30, I noticed a post written by Todd, a guy I'd worked with at Keystone. One of those drunken posts of a type I might have written once or twice. I hadn't kept up with Todd much, though I thought I remembered seeing some post about a trip to Rome he took with his wife and kids not long ago. This fresh one had been written only a minute earlier, and it was downright bleak. He'd been diagnosed with bladder cancer, he said.

Todd wrote in his post about how he'd had no luck finding work since leaving Keystone. His wife had gone back to teaching for the insurance. Their credit cards were maxed. They were trying to shelter the kids, but they were barreling towards the worst Christmas ever because they couldn't afford gifts. *You think once you've made it, you've made it*, he said. *Then it all gets unmade*. It was sappy and full of misspellings. They were two mortgage payments behind and might soon have to move in with his in-laws. Jesus.

I liked Todd. He was a good guy. Reading his late-night post bugged me. Sam and I had been through our own share of less than stellar Christmases, and I remember feeling like shit about them at the time, even if she never showed her disappointment.

I scrolled on.

Luis had posted a picture of one of his cats who'd died the year before. "A whole year since my good buddy Spicolli died," it read. Some people he sold coke to had liked the picture.

My mother had posted a link to an article about a left-wing politician embroiled in a scandal and wrote, *If Trump did this he'd be in prison.*

I came across a video from an account that always posted sentimental videos about how sad it is for a father to see his daughter grow up and move out. I reposted it and tagged Sam.

Two people I knew posted that their accounts had been hacked.

My mind kept returning to Todd and wondering what he would do. He was a bright guy. He seemed like a kid to me, even though he was in his early 40s. When you're young, you can't imagine ever thinking of forty-three as young, but now I did. He was thirteen years younger than me. I kept going back to the post and rereading it. So pathetic and vulnerable. I'd scroll on and then return to it. Scroll and return. Finally, I went back to it and found that it was gone. I went to his profile to read it again and saw he had deleted it. I was probably the only person in the world who had seen his little message in a bottle.

I put down the phone, but still couldn't sleep. Tonya was sawing logs beside me. God, I didn't envy her. A divorce. She'd said before she wanted my advice. Maybe I was getting off easy just lending her some money. What difference would five thousand make to me now?

I got out of bed without waking her and went to the closet. I opened one of the safes and took out two ten thousand-dollar bundles, broke the band on one of them and counted out five thousand on the bathroom counter. In my office I found two yellow envelopes. I put fifteen thousand in one and wrote "Todd" on it. I put five in the other envelope and wrote "Tonya" on it. I set them both on the entryway table and went to bed. Finally, I slept.

When I awoke the next afternoon, Tonya was gone. I ordered lunch around two, went to the front door to retrieve the bag, and found that Tonya's envelope was gone. Until seeing the remaining envelope marked "Todd," I'd forgotten all about the money and his Facebook post. Jesus, another errand.

# 30

TODD'S HOUSE WAS in a neighborhood that was once solidly middle class and was now solidly unaffordable.

Like most businessmen in Texas, Todd drove an F-250. And like most wives of Texas businessmen, his wife drove a leased Mercedes. Younger wives went for the SUV, empty nesters went for the coupe. Todd's wife had the SUV in silver.

I hadn't called or texted to warn him of my visit. I didn't know why, but I wanted to catch him off guard, to see him in person and get a measure of his misery. I could see that a late-night version of myself had made a plan the daytime version of me didn't quite understand. But the late-night version of me had been right before.

Todd answered the door barefoot, wearing jeans and a navy-blue polo.

"Winston! How are you?"

"Not too bad," I said. "How are you?"

"Not bad," he said, still keeping up the old cheeriness. "Can't complain."

"I want to take you out to dinner. Tell Amy you're going out."

"Is everything okay? She's making spaghetti. You should come in. You could eat with us."

"No, I'd rather go somewhere," I said. "I was thinking Bogart. My treat."[1]

"Bogart sounds good," he said. "I haven't been there in a while. But at least come in and say hi to Amy."

"No, thanks. I'll wait out here. Tell her I said hi."

He came back out wearing a pair of brown loafers and a sports coat over his polo, and we were on our way.

"You're looking good," he said.

"I look fat."

"Healthy," he said. "My mother always says people look healthy when they put on a few pounds."

"I wish your mother would tell that to my mother. None of my suits fit anymore. By the time the noncompete expires, I'll have to get a whole new wardrobe."

"Your noncompete? You're still hung up in that? How is that possible?"

"Ask Bryce."

"What a piece of shit."

"You know the fucker still calls me sometimes?"

---

[1] Bogart was a steakhouse we'd been to dozens of times. We'd done wealth management seminars in their meeting room. It was how Keystone got about sixty percent of their clients, and I was always the presenter. I was extremely good at it. I remember reading that most people rank their fear of public speaking just above their fear of death. I never had much fear of either, which I guess made me ideal for the job. For a time, Todd had trained under me as a replacement rainmaker for the nights when I didn't feel like doing the whole seminar song and dance. His conversion rate was about a third of mine, but it was still impressive by most measures. I hadn't been to Bogart since March of 2020 when the first shutdowns led us to cancel one of our seminars.

"What for?"

"Fuck if I know. Wants me to help clean up messes at Keystone. Messes that have literally nothing to do with me."

"And he still won't let you out of the noncompete?"

"Nope."

"What do you have left? Six months?"

"Ten."

Todd shook his head. "You and me out here struggling, and fucking Bryce building his new house. Have you seen the pictures online?"

"No, I have him blocked."

"This guy is building a fucking palace. As if the other one wasn't big enough."

The bartender, a dude in his middle forties named Jacob, recognized Todd and me the moment we walked in the door. We had once been real VIPs there, and Jacob was a terrific bartender. He even remembered Sam's name and thought to ask about her. Though I was sure he remembered the name of the last chick I had dated, he was tactful and smart enough not to ask about her. When we were seated, Jacob sent over a waiter with two shots of tequila.

"On the house," the waiter said, displaying a bottle that looked like something out of *I Dream of Jeannie*. "Two hundred dollars an ounce, usually."

We drank and nodded our approval. "Not bad," I said.

He offered us menus, but we didn't need them. We both ordered the filet, bloody.

"I saw your post last night," I said. "How sick are you?"

Todd's face changed. "Oh," he said. "You saw that? That's right. You're a night owl. I was just having a moment. I'm not that sick. I'll have surgery next month. We caught it early. Sorry. I shouldn't have put that shit out there. I drank too much."

I took the envelope from my jacket pocket and put it on the table in front of him.

"What's this?" Todd said.

"Fifteen thousand. You gotta pay it back, but you don't have to be in a hurry about it."

His face shifted again. He put his hand on the envelope and began to slide it back towards me.

I stopped its progress. "It's not a discussion."

I could see the emotion rising in his neck, turning it red, tightening its way up to his tear ducts.

"I'm going to go take a leak," I said.

I went to the bathroom and pissed in one of two urinals where I'd pissed a thousand times. Then I went to the bar to thank Jacob for the tequila and stood listening to him describe its provenance.

"I should get back to Todd," I said at last.

"It's very good to see you again, sir," Jacob said. "Hope you guys will get back to it soon."

"We'll see. Thanks again."

When I returned to the table, the envelope was gone, and a bulge stood out in the breast pocket of Todd's jacket.

"You know, I think I have a lead on a job for you if you could get out of the noncompete," he said.

"Oh, yeah? With who?"

"Arnold Hershey."

"At Trinity? How are they doing?"

"Oh man, they came out of Covid smelling like a rose. I had lunch with him a few weeks ago. He asked about you."

"Are you going to work there?"

"I don't know. I need to figure out the cancer situation first. Amy wants me to take it easy until after the surgery. Doctor said I should be able to go back to work soon after. We'll see."

"I don't know Hershey very well," I said. "He seems a little straight. What did you think of him?"

"Definitely very straight. It was a dinner meeting. His wife was there. We had to pray before we dug in."

"At a restaurant?"

"Yeah, but he makes a lot of money. A lot of these born-again Christian types got real antsy about the end of the world and all with Covid and whatever the fuck is happening. Arnold has the magic touch with them. He uses all these faith-based funds and stuff. Some of them aren't as bad as you would expect."

"I don't figure a guy like that would be too interested in a guy like me."

"He likes to reform people, I think. So, it's kind of like the worse the better. When he was in the bathroom, his wife asked if I had accepted Christ as my personal Lord and Savior."

"What did you say?"

"I said of course."

"Yeah, that's what I'd say."

"Maybe Hershey will still be interested in reforming me in ten months when I'm out of the noncompete."

"Surely you can get out of it before then. Seems so fucked up you're the only one who got two years."

"I was the only one dumb enough to sign it."

"You were the only one valuable enough that they even bothered making it two years."

"That's one way of looking at it."

Before I dropped Todd off at his house, I asked him not to tell anyone other than his wife about the loan, and he agreed. He also managed to drop into conversation the precise location of Bryce's new house.

"Maybe you could drive out there sometime. It's real secluded. They just finished framing it."

Todd knew me pretty well.

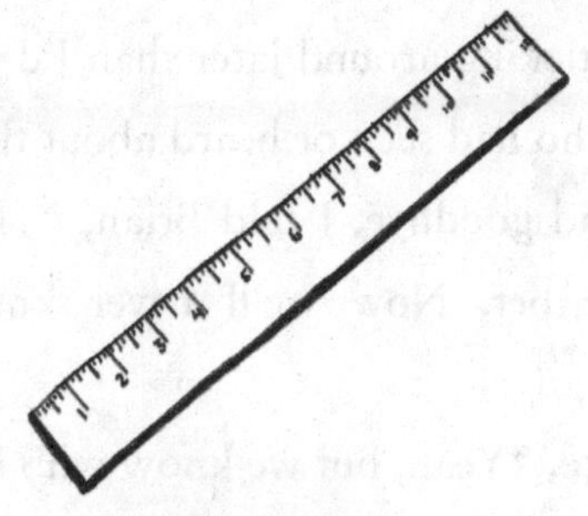

# THE DICK MEASURING CONTEST

I remember going to a party with some people I knew through my music friends. Some loudmouth was talking about this chick I'd met at a show. He'd fucked her recently, and he kept saying she was blown away by how big his dick was. She told him it was the biggest dick she'd ever seen in person. He was obviously getting off bragging to the ten or twelve people in earshot.

"Bet my dick is bigger than yours," I said.

The guy kind of laughed. "Anything's possible."

"How's this sound?" I said. "We both take out our dicks and let everyone here decide. I bet you $5,000 my dick is bigger than yours."

"Yeah, right."

I held his gaze in complete seriousness.

My buddy Brian was there. He said, "Wince is good for it."

It was true I was. I always paid when I lost a bet.

The guy debated it for a while, but he ended up blowing me off, acting like the whole thing was a joke, even though it obviously wasn't, and he left the party not long after.

I ended up sticking around later than I'd planned, mingling with the folks who had seen or heard about the bet.

When we said goodbye, I told Brian, "That dipshit should have taken the bet. Now we'll never know who has the bigger dick."

Brian was like, "Yeah, but we know who has bigger balls."

# 31

THE NEXT WEEK was Christmas. I told my mother I wasn't well enough to make the drive, and when she suggested they might come to my place and take care of me for a week or so, I texted her a picture of the positive Covid test I'd saved.

*Still contagious, as far as I can tell*, I said.

While everyone else was celebrating, I was in my garage with a router and a stack of AR-15s. Sometime that night, my daughter unwrapped a gift under the tree at the house of her boyfriend's parents, and I received the call I'd been expecting for weeks. They were engaged, and she was ecstatic.

"We're going to do everything in the traditional way," Sam said. "Daniel's parents are very traditional. It's all going to be by the book."

Traditional. She'd said it twice. I guessed that meant her future in-laws were saying it. And I guess I knew what they meant by it, too. Traditional was code for the-father-of-the-bride-pays-for-the-wedding. Whatever Sam wanted. That's what I told her.

"I'm so happy for you, baby," I said. Of course, I was crying. "What an incredible Christmas gift."

But I wondered why, if they were all so traditional, Daniel had forgotten to ask me for her hand in marriage.

Next came the calls from my parents. My mother had just gotten off the phone with Sam and was reading aloud from a forty-year-old World Book Encyclopedia set. Facts about Venezuela, whose primary exports were petroleum and natural gas, with an elevation of etcetera, etcetera. I let her go on as I made my way to the kitchen, muted the phone, and did a couple of lines.

"It doesn't say anything about their wedding customs," she said.

"Pretty sure they cut the head off a live chicken."

"Oh, God. Surely not!"

Later, I received separate texts from Todd and his wife. Pictures of their joyous kids, a boy and a girl ecstatic over their plentitude of gifts.

*So much gratitude*, Amy wrote. *You're our hero.*

Timothy, my old assistant, texted to say Merry Christmas.

Luis sent a Santa emoji with the words, *Feliz Navidad, hombre. When you coming over?*

I heard from Bryce. *Merry Christmas, old chap. Hope it's a merry one.* A half hour later, he texted again. *I have a gift for you. Nothing big.*

I didn't reply to him.

I worked on the guns, thinking of tethering Bryce to the workbench and beating him with an oversized wrench someone had once given me as a joke. It would be perfect for shattering his jaw.

When I finished working, I texted him a picture of my father's boat in Arkansas that I'd taken over Thanksgiving. *Out of town*, I said.

*When you get back then.*

I went to Luis's house. We smoked one of his killer joints, and he told me what he knew of the latest run to the junkyard. Since I had turned him down, Hector took Luis's son Manny, who had accidentally discharged his gun into the seat of Hector's Bronco. Hector hadn't let Manny or Luis hear the end of it. Luis laughed telling the story. Fuckin' Manny. I was glad it hadn't been me. Of course, it never would have been.

Luis tried persuading me to go to a family dinner with him, but the thought of being in a room full of his kids and grandkids was too much, so I paid for my coke, did a line with him, and hit the road.

"Sure you can't come?" he said. "No one wants to be alone on Christmas."

But I did.

I'd loaded two chainsaws in the back of the Tahoe before leaving the house. I did bumps at the stoplights on the way out to Bryce's new property, which I'd looked up on Google Maps that afternoon. The roads were practically empty. Everyone was at home unwrapping gifts. How Bryce managed to find such remote properties was anyone's guess, but he had succeeded once again. The house was up a long, muddy driveway newly rutted by construction trucks.

I got out and looked the place over in the headlights. Todd was right. The house Bryce was framing made my parents' house look like a guest cottage. I walked through it, imagining him in the various sprawling rooms, most of them with incredible views of the lake. Trees filled every view that wasn't taken by the water. They were everywhere, a tight little woods pressing in on all sides. Walking through the hollow house, smelling raw lumber, I pulled my collar closer around me and listened to the air blowing, whistling and knocking around beer cans left by the framing crew. It always made me miss construction being in a house in that state. I could remember every framing hammer I ever owned. Sometimes I wished I could go back to those days, with sawdust in my hair.

I could see the house as it would look when it was complete, and I was, of course, filled with envy. Todd told me Bryce had a scheme to get a farm exemption on the land to keep his property taxes at next to nothing. He'd raise bees, away from the house where no one would be bothered. Even that irritated me. Not that I gave a shit about everyone paying their fair share or anything. But why should Bryce get to skirt the rules when he was such an absolute piece of shit?

I went down to the Tahoe and took out one of the chainsaws I'd packed. I hesitated a moment, worrying over the sound in the quiet Christmas night, but I figured if anyone across the lake heard, they wouldn't be able to tell where it came from. I took it as a good omen when the chainsaw started on the first pull. I fiddled with the choke, tuning the saw like an instrument until it sounded the way I wanted it to.

I figured if I took out enough of the exterior studs, along with the corner posts, the house would fall in on itself completely.

However it collapsed—partially or fully or whatever—they'd have to start the work all over. And they'd have to go through the giant pain in the ass of taking apart a big wad of boards and hauling it off. I walked to the nearest corner, depressed the trigger and began.

After slicing through most of the exterior walls, I stepped back and watched the two-by-fours buckling and wrenching against their nails. The house broke inward like a piece of paper being wadded by a pair of invisible hands. The noise of its collapse was even louder than the sawing. So loud it startled me and almost sent me running to the Tahoe. But after it fell, it was quiet again, and I wanted to linger there a while longer looking at it. It was beautiful. Watching that bastard fall was the only Christmas gift I got. I stood for at least ten minutes, with the cold wind making my ears feel like they might bleed, savoring it, feeling like maybe my days of doing nothing might be behind me.

# 32

WHEN TONYA SHOWED up at my place late Christmas night, I was pretty fucked up. Celebrating the destruction of Bryce's new house, I'd inhaled most of the coke I'd gotten from Luis and had gone slightly too far.

"Jesus Christ, you're red!" Tonya said when I answered the door. "You look like a Christmas ornament."

"I'm fine."

But the truth was I was relieved to see her. My heart was feeling shaky, and I found myself wishing I could take back those last few lines.

"You look terrible, Winston. Go look at yourself in the mirror. You're sweating. You're drenched. And you stink."

"I do not."

She leaned in and pressed her nose to my chest. "You do. Like body odor. And sawdust and gasoline. The fuck have you been doing?"

"Nothing."

"Honey, let's get you into a shower. Seriously, I'm worried."

"What are you worried about? I'm fine."

Seeing how anxiously she looked at me intensified the unsteady feeling I was already having.

"Do you have a blood pressure monitor?" she said.

"You've got to be kidding me."

"Please, let's get you in the shower. Or something else. What makes you feel better when you're in this state?"

"I don't know. Weed?"

"I don't think you should get any more fucked up in case we have to go to the hospital."

"We're not going to the hospital, Tonya. Stop it."

"Well, go look in the mirror. You look—"

"Stop telling me how I look. Calm down. I'll pour us a glass of wine, and we can sit and watch some TV, alright? I agree I overdid it. Now please let's drop it. I'll get in the shower in a minute. I just need to chill out a little bit first. You're doing the opposite of calming me down."

"Okay," she said. "I'm sorry. You're right. I'm sure you're fine. But we should find a way to bring you down a little."

We sat on the sofa watching TV, and I reached to hold her hand. She gripped my sweaty mitt without saying another word about my appearance or smell. I debated pausing *Reindeer Games* to tell her everything, how it felt to watch that house fall exactly like I had hoped it would.

By then, she'd spent months listening to stories about what a piece of shit Bryce was. But I still couldn't predict what her reaction would be. So, instead, I told her about my daughter's engagement, and my mother's concerns over our cultural differences, and Hector's trip out to the junkyard with Manny, the

Annie Oakley of our time. Eventually, I felt myself calming down, and, after a trip to the bathroom, I debated a detour to the kitchen for a final bump of the night.

"Don't even think about snorting anything," Tonya said. "You've got to be fucking kidding me."

"Jesus, would you relax? I'm getting another glass of wine."

Eventually, we went to bed, and Tonya stayed until I fell asleep.

"Thanks for the loan, by the way," she whispered. "I'm going to pay you back."

"I'm not worried about it," I said.

When I awoke the next day, she was gone, and I could smell the stink of myself in the sheets.

I had only one more holiday to dread.

# ROLLING OVER

I remember the one time I had to roll over on an associate to stay out of trouble. I was alone, flying a borrowed airplane full of coke to a little regional, non-towered airport where I'd brought in shipments at least a dozen times with no problem. I didn't have a pilot's license, and, though I didn't consider it stolen, I didn't exactly have permission to use the plane I was flying either.

As I got close to landing, I noticed three tan Crown Victorias and a black Suburban parked at the airfield, where there should have been exactly one car—the white Honda Accord which had been provided to me to make the delivery. I debated the landing, decided against it, and quickly pulled up. The Crown Vics took off, doing their best to stay with me from the ground as I made my way to the next nearest airfield, which was the only place I could possibly land, given my limited supply of gas.

There was no way they could follow me the whole way, but they must have guessed what I had in mind, because when I arrived at the airfield, two more sedans were already waiting. I landed, was tossed into someone's backseat, and told that if I made the delivery as planned they'd let me off. I didn't even have time to call a lawyer. I signed the document they shoved in front of me and gave them the keys to the white Accord parked at the first airport. Their agents gassed up the stolen plane and flew out to get the car while they interrogated me. There wasn't much for me to tell. Though I had brought plenty of coke through that airport, on this particular occasion I was working with guys I wasn't that familiar with, which was probably why I got in trouble in the first place.

In the end, they brought me the car, helped me load forty kilos of coke into the trunk, and followed me on a chain-smoking drive to the meeting spot at the house of someone I'd only met once before. When I backed the car into the driveway, the two Crown Vics trailing me stopped at the corner while I went inside.

I said hello to the man I was supposed to meet—a white guy in his forties with a Fu Manchu—and apologized for my tardiness. He asked if I needed a ride somewhere, and I said I didn't, that a friend of mine lived around the corner. I walked out the front door and down the block where a man in a suit picked me up and asked where I wanted to be dropped off. By then, they knew everything about me so, after a moment of hesitation, I gave the man my home address.

The one thing I remember saying to the agent who drove me home was, "What's going to happen to that guy?"

"He'll do the same thing as you," he said.

It eased my guilt imagining that the guy I'd rolled over on would roll over on the next guy. I remember telling myself I would never have done it if I had known him personally, if we had been friends. But how many more visits would it have taken for me to consider him a friend? Maybe my whole life would have been different, and I would have gone to prison if we had gone out to lunch once or twice.

## 33

THE WEEK BETWEEN Christmas and New Year's, I went out to Hector's land to test the guns while the maids were at the house.

I texted him first.

*Hey, headed out to your property. You still cool with that?*

He called instead of texting. I hated when people did that.

"Hey, man, yeah, the land is good. You're heading out there now?"

"That's the plan."

"You got 'em all ready? Ten?"

"Pretty close. We'll see."

"I'd come out to meet you, but I'm tied up."

"That's alright. Hopefully, it'll be a pretty quick trip. I got other stuff I gotta do today."

"Alright, man. So, maybe we'll hook up later this evening at my dad's place."

Hector's land was exactly what I'd hoped: private, remote, and not too far from the city. He had a scrubby little hillside perfect for testing the guns. He'd obviously used it for the same purpose. There was a lot of brass on the ground. I didn't bother setting up targets.

Every time I tested one, I held my breath, but all ten worked flawlessly, thank God. I texted Hector to tell him the guns were ready, and we set a time to meet at Luis's later that night.

While I was there, Todd texted. Since I lent him the cash, he'd been reaching out more.

*Hear what happened to Bryce's new house?*

*No*, I replied. *Just got back into town. Been visiting my folks. What's up?*

*It's destroyed.*

*What happened?*

I almost felt I might swoon reading about my acts secondhand. I wished he could have sent pictures. I couldn't stop thinking about it, about how I'd really done something, taken some kind of action. When had I last felt that good?

In fact, ever since the night at Bryce's property, I'd been thinking about what might come next. The house couldn't possibly be the end of it.

On the way home, I called Boy Wonder.

"Hey, boss," he said.

"You at work?"

"It's Saturday."

"Oh, well that's good then. Hey, that time we went to Keystone and fixed that stuff for Bryce, you've done that sort of thing for them a lot, right? When we were working for them?"

"Sure."

"Well, could we report him to someone without getting ourselves in trouble? Like for whatever sleazy shit they've been doing?"

"I guess we could," he said. "I mean, FINRA, for starters. And the SEC, the National Association of Insurance Commissioners, the Texas Department of Insurance. I could probably think of a couple more. You thinking of threatening him?"

"I'm past threatening. You know how to file a complaint?"

"No, but it can't be that hard."

"Well, work on it, would you? It could be a little side project for you. Confidentially. I won't tell anyone about your involvement."

"It might be a little time consuming," Timothy said.

"How about fifteen hundred?"

"Per complaint?"

"If we're talking per complaint, let's say a thousand."

"No sweat."

"Start with two. FINRA and the Texas Department of Insurance. I'll Venmo you a thousand right now so you can get started this weekend."

"You want it anonymous, I'm assuming."

"No, I want him to know who did it. I'm going to send him a copy when it's done."

"As long as it's your name on the paperwork and not mine."

My whole life I thought the best thing you could possibly be is a criminal with a code. All the fun and excitement of being a bad guy, doing whatever the hell you wanted, but in a deeper sense, where it mattered, you were a good guy, following the code. There was one rule every crook had heard a million times: don't snitch. You've got a problem with a guy, you fight him, burn his house to the ground, fuck his wife, but don't involve the cops.

Other than that one time with the plane, I had never done it. And even then, I hadn't known the guy. It was shitty, but at least he was a stranger. I think in some stupid way, I must have still thought of Bryce as a friend, and that's what had kept me from taking any real action against him. I kept accidentally remembering him as he was before he had risen to his place of prominence at Keystone, before his ambitious clawing had paid off. However much I hated him and dreamed of torturing him, the other memories were still mixed in. It had taken me this long to understand that snitching on Bryce would violate my code far less than rolling over on a coke dealer I never knew.

Whatever happened to Bryce, not in my mind but in reality, I hoped it would be horrible. The worse the better. If it weren't for him, I wouldn't be worrying about the cost of lending some cash to my friends or paying for my daughter's wedding. I wouldn't be driving out to bumfuck Egypt to test firearms that could get me a lifetime in prison.

When I pulled into my driveway, the yard guys were at work on the lawn. I didn't think to hide until it was too late.

## 34

THE SITUATION WITH the yard guys had been going on since the day I started renting the house, five or six years earlier. When the moving truck was still in the driveway, I saw a crew of guys mowing the lawn next door. My lawn needed mowing too, so I asked the guy in charge how much it would cost to have them give my grass a quick trim. He said fifty bucks but agreed to do it for thirty. I gave him the cash, which I'm sure he pocketed, and his crew did the work.

Without me ever saying anything else, it got into the heads of the guys who mowed the neighbor's lawn that I, too, was one of their accounts. From then on, like clockwork, they came to mow my lawn every other week even though I never paid them. As the months passed, I understood what had happened, but I kept my mouth shut. I always figured I had plausible deniability since they had done it from the moment I moved in. I could always say I assumed my landlady paid for the service. Still, I was smart enough to hide when the lawn guys were at the house.

They must have done some kind of internal audit of their accounts, because about a year earlier, I started getting letters requesting payment for services rendered. I ignored them, obviously. Still, the guys continued mowing my lawn every other week, and I continued hiding out. It escalated to the point where every time they mowed, a crew leader would come knock on my door. Then, somehow, they got my cell phone number. That really surprised me. Sometimes they left messages, and the one time I answered by mistake, I told them they had the wrong number.

That afternoon, having fired ten machine guns and plotted the early stages of destroying the career of Bryce Winters, I was sufficiently distracted that I let down my guard and failed to notice a dignified-looking Mexican wearing khaki pants and a grey polo standing on my front porch.

"You're Mr. Fisher?" the man said.

He thrust out his hand for me to shake and introduced himself as Pedro Padilla, informing me that he was the owner of the lawn care company that serviced my yard.

I was dumbstruck, but it didn't matter, because he charged ahead, explaining in exceedingly polite terms that I had a delinquent account with his company, and I owed him roughly seven thousand dollars.

"I don't see how that could be right," I said. "My landlady pays for the service. She's had my lawn mowed the whole time I've lived here."

"I wondered about that, but I contacted the owner," Pedro Padilla said, "and she sent me a copy of the lease where it says that the renter is the one responsible for lawn care and maintenance."

"Well, I know I never signed a contract with your company. This is obviously an error on your crew's part."

"Twice a month for almost six years is one hundred and thirty times we mowed your lawn. We charge fifty dollars for a lawn this size. Even if we say it was only five years, that's sixty-five hundred dollars."

"I never agreed to fifty dollars," I said. "I never agreed to this service at all. I know for sure I never signed a contract or made any verbal agreement with you. And, incidentally, if I never hired your guys, and my landlady never hired your guys, they're trespassing every time they come into my backyard."

"You seem like a nice guy," Pedro Padilla said, smiling warmly. "A very nice guy." He spread his hands. "And to live in a nice house like this, I'm sure you understand business. This is my business, sir. I'm happy to mow your lawn. It is very convenient for us because we have other clients on your block. We charge fifty for every lawn. We rake your leaves and grass clippings. We do all the edging. We try to do very good work for you. Have you been happy with the work we've done for you?"

"It's been fine," I said. "But, still, there's no way I could or would pay anything close to seven thousand dollars."

"I know what you mean," he said. "But you do owe us that money. I'd hate to have to go to the courthouse and talk about all this in front of a judge. I'd rather be outside with my crews on a day like this, talking to my customers who seem like nice people. So, what do you think we should do?"

I sighed and looked at the lawn. "There's no way I'd pay you more than like two thousand dollars, man, and that would be to wipe the slate clean."

He smiled and held out his hand. "Deal."

How could you not like a man like Pedro Padilla? It was all done so smoothly and with such kindness and respect.

Before he left, as his men mowed my lawn, I went to the safe and withdrew from the dwindling stacks another two thousand dollars in hundred-dollar bills, wondering how I would have gone about collecting a debt from someone who had no intention of paying. I knew from the past that my way would have involved more threats, more anger. It was a good reminder that there were different ways of doing things.

Mr. Padilla, receiving the money, held it in his hands as if he had received a great blessing.

"This way, I will only have to listen to my wife yell a little bit when I get home tonight," he said. "And, of course, we will be glad to continue to service your lawn if you pay us monthly. Why don't you give me your email address so we can bill you that way moving forward? You can pay through our portal. It's very simple."

Fucking portals. Everyone had one now.

Before he left, I said, "You do all this shit, confronting people about overdue bills and stuff, without a gun?"

"No, sir," he said. "I have a gun. If you ever need to get in touch with us, you can call or text me anytime."

I sat on the front porch having a cigarette while his crew finished up—at least I didn't have to avoid them anymore—then got into bed with the sound machine playing ocean waves, dozing until it was time to meet Hector.

I got to Luis's place early and, as we got stoned, he told me about the nurses at the dialysis clinic where now he spent three afternoons a week. He was already selling coke to two of them.

Hector looked upbeat when he breezed through the front door. He sat and visited with his father and me, then the two of us went out to the garage. I had backed the Tahoe up to the door so I could easily unload the guns. Hector looked at them laid out on a blue tarp we'd spread over the grease-stained floor.

"Not bad."

Seeing them there, I almost couldn't remember having made them, having done all the steps ten different times. No part of the process stood out. Even firing them earlier in the day, one after the other. It was work now. I had made a job for myself.

Hector asked about my Christmas, and I about his, then he gave me the cash in a small, green canvas tool bag. $80,000.

"When can I get the next batch?" he said.

"How many more do you want?"

"Could you do twenty?"

"No," I said. "I don't think I could. I've got a lot of other stuff going on right now, and I have some parts I need to get my hands on."

"How many could you do?"

"Five," I said. "But I'll need at least two weeks. The supply chain stuff is fucking everything up."

"Maybe I could ask someone I know to help you track down what you need."

"I'm not looking to involve anyone else."

"Okay. Well, more would be better."

"That's all I can do for now. Covid slowed me down a little. I'm still recovering."

"Yeah, man. I get that, but think about it. I've got other sources for parts if yours are drying up."

"Got it," I said. "I'll think about it."

I went inside to say goodbye to Luis and told him I'd be back soon.

"Next time I'm going to show you the new *Mission Impossible*," he said. "You'll love it. Lots of action."

"Sounds good," I said, though we had already discussed that I'd seen it and hadn't thought much of it.

There was something I didn't like about the pressure Hector was putting me under. However much I liked the guy, I didn't start making machine guns to have another boss telling me what to do.

When I got home, I carried the bag of cash to the kitchen, did a bump off the counter, and noticed I could hear the washing machine running. Hadn't it been running when I left? I looked into the laundry room and found the washer stuck in an endless rinse cycle. I tried everything I could to reset the machine, hitting all the buttons and even unplugging it and plugging it back in, but nothing would make it drain or enter a new cycle. I had bought the pair of machines a few years earlier, just long enough for them to be out of warranty.

I texted the housekeeper. *Hey, did you notice something is wrong with my washing machine?*

*Yes, it's not working*, came the reply.

"Well, no fucking shit," I said to the washing machine.

This kind of housekeeping idiocy had become utterly routine. The two housekeepers had been with me longer than any of my three wives, but they were absolute morons who fucked up more at my house than they corrected. My list of complaints against them was essentially endless, a fact which Tonya found hilarious, despite my actual, genuine frustration over them. They broke things, misplaced things, refused to use the dishwasher, never told me when we were out of cleaning supplies, and shrunk my clothes.

It had once seemed paranoid, but now I believed they were straight up fucking with me. The week before, I had been unable to find my reading glasses, and when I texted to ask if the housekeeper knew where they might be, she replied two hours later. *Under the sink*. And there in the cabinet beneath one of the two master bath sinks were my eyeglasses, behind a bottle of mouthwash.

I felt bound to them after all those years, not due to any sense of loyalty, but because they were so accustomed to seeing drugs and guns everywhere and, though I was certain they pinched some of my weed practically every time they visited, from what I could tell, they left my guns and money alone. I couldn't exactly Google housekeepers who didn't mind coke, weed, guns, and the possibility of finding me passed out naked in bed.

As I stood over the endless agitation of the washing machine, watching YouTube videos about methods for resetting it, my phone pinged with another message from the housekeeper.

*Been meaning to tell you. Now we charge $250. Should we keep coming?*

This fucking bitch. It was double what I had been paying. I didn't even ask them to touch the upstairs. Through the entire first six months of Covid, she wouldn't come to my house for fear of getting sick, and I continued paying her usual rate. Now she was doubling my fucking prices. Finally, I replied. *Yeah, we should talk but keep coming for now.*

I went to the kitchen and did another bump, then carried the bag of money to my closet where I unloaded the eighty grand. Standing before my safe and looking at the stacks of money, I recalled that a voicemail had come through while I was out shooting that afternoon. It was from my landlady. I didn't need to listen to the message to know she was jacking up my rent. It wasn't until I got good and stoned that I could bear to listen to the message to verify I was correct. I fucking knew it. Another seven hundred and fifty dollars a month. It was still probably cheaper than almost any other rental house in the area, but it was more than I wanted to spend on a place that was too big for me by half.

The hits kept coming. That night, I sat on the couch watching commercial-less episodes of *Two and a Half Men* on a Hulu account my last girlfriend had never signed out of. I watched one after another. My phone pinged.

*Daddy, what are you thinking about the budget for the wedding?*

I ignored the text for a half hour, which is probably the longest I'd ever deliberately avoided a text from Sam. Finally, I replied. *No idea. You?*

*Amanda's getting married next January, and we have basically the same friends. She's already done a lot of research. She's trying to do everything on a budget, and she's at $45K.*

Jesus Christ, I hadn't spent forty-five thousand dollars on all three of my weddings combined.

*Seems high.*

*If we keep it small, we could prob do $40. That would be okay?*

I checked my accounts on my phone. Every stock I owned was down. No exaggeration. Every single holding was in the red. I couldn't remember the last time there was even a hint of green.

*40 is ok*, I said, hoping by my brevity she would surmise that I meant the opposite. *But less would be better.*

*Thank you, daddy. We'll try to keep it in that range.*

With my fingers on top of my grandfather's table, I tried counting all the people who wanted money from me.

Sam for her wedding.

Luis for his drugs.

Tonya for her divorce.

Pedro Padilla for his lawnmowing.

Timothy for the SEC complaints.

Todd for his cancer.

The housekeepers to hide my glasses.

The landlady for her shitty, oversized house.

God only knew what a new washing machine would set me back.

My eyes travelled to the spot next to my television where I'd hung the picture of the junkyard. It had been there for weeks while I watched TV. Hours of movies about robberies and heists. They always went wrong. It got you thinking down that path a lot, what kinds of things could throw a wrench in a project like that. That's how I thought of it: a project. The guns had been a project, and I had made some money. But not enough.

The last few junkyard pickups Hector had made, he'd made on Wednesday nights. Not too early and not too late. Realistically, how hard could it be to go to the junkyard, take out the guards, get into the room where they kept the coke, kill whoever was in there, and take everything? I could tell from the firearms they carried they didn't know shit about shit.

It was like anything else. It seemed like a big pain in the ass in the abstract, but usually it was just a matter of getting off the couch. I'd stolen from drug dealers before, and I'd always survived. Though, admittedly, that had been decades earlier, and I had been shot once and hit with an axe another time.

I took down the map I'd drawn and put it on my grandfather's table. I watched TV and thought about it and didn't think about it. That's the way it works. Making a plan is like remembering a name you've forgotten. You think about it, then you stop thinking about it. Then you think about it, and you stop thinking about it. When you stop thinking about it long enough, your mind starts opening all the doors in a corridor until it finds the answer.

# WING SHOT

I remember a time when practically everyone I hung out with was a biker. It was a pretty fun life, the way my buddies and I could take over traffic, take over a store. A big group like that could dominate the vibe pretty much anywhere we went. But even alone, we all got treated with this kind of grudging respect because people were afraid of us. This was back before Harley culture transformed into this bullshit dentist-and-lawyer scene.

One night, I was with friends at this biker bar when I got into it with another guy about Harley engines, of all things. As if anyone could out-argue me on that subject. The argument turned into a scuffle, and the guy pulled out a little .32 and shot me in the arm as I was mid-swing.

The thing no one tells you about getting shot is how much it burns. You've got this scorching piece of metal in your body, searing flesh, blood vessels, and muscle. The bullet cools down on its own time, which isn't nearly as fast as you'd like.

The fucking thing broke my arm. Ruck drove me to the hospital, and the asshole who shot me ended up in prison. Like with everything else, I still have the scar.

## 35

BOY WONDER CALLED. He was a machine. Timothy had already hammered out the FINRA and Texas Department of Insurance complaints against Bryce.

"I emailed them to you," he said. "Can you get to your laptop?"

I opened the first PDF and skimmed it before moving on to the second. They were brilliant. Of course they were. "This some kind of prescribed format you're supposed to use?"

"No. They don't give you much guidance. You sort of complain however you want."

"If I save these files and send them, is there anything linking them to you?"

"Nothing they'd bother digging into. I sent you the email addresses of where to send them. I'd use a subject line like 'Urgent: suspected ongoing securities fraud.'"

"Hang on, I'll send the first one now. Which one do you think would scare him more?"

Timothy thought about it. "Well, what's your aim?"

"To send a complaint to every agency I can think of, a new agency every week until he lets me out of my noncompete."

"Then start with the Texas Department of Insurance. Everyone's afraid of Texas. So, what happens if you're free of the noncompete? Are we going to start our thing?"

"Hell, I don't know, Tim. I'm barely putting one foot in front of the other."

"Alright, boss. Let me know."

I wasn't even sure anymore why I wanted out of the noncompete so badly except to put the screws to Bryce. It was hard thinking of myself as the same person, with fifteen appointments on my calendar every day. I wasn't sure I could ever go back to that life.

I texted Tonya but didn't hear back from her. I wanted to talk to her about the situation with Hector. I'd barely gotten the last ten guns completed. Now he wanted twenty, and I'd agreed to five. I could probably scrounge together what I'd need for twenty builds if I asked a few people to help me get the parts, but it would be a huge pain in the ass, and not everyone would be as incurious as my father. I told myself I'd keep thinking about it and not thinking about it until an answer bubbled up.

The junkyard was right there on the table.

# 36

Two days after I sent the first complaint to the Texas Department of Insurance, I emailed Bryce.

*Bryce, I am writing to let you know that every week you keep me in this noncompete, I'm going to file a complaint with a different regulator. I submitted the first one on Monday to the Texas Department of Insurance, as you can see in the attached PDF. I have another one written and ready to send to FINRA, which will go out on Monday. After that, I'll send complaints to the SEC and the National Association of Insurance Commissioners. When I run out of places to complain to, I'll start following up with each agency to prevent a potential financial criminal from slipping through the cracks. I'll be free to discuss the matter one-on-one for one week only. After that, I will send you the contact information for my attorney.*

Fifteen minutes after I hit "send," my phone rang. Stein, Frank N. I let it go to voicemail. The phone rang again. Good. Let him suffer for a while.

How could it have taken me so long to turn on the bastard?

A text came.

*Hey, call me.*

Another one arrived five minutes later. Another an hour later.

Finally, I texted. *What's up?*

*Saw your email. Can we talk?*

*Can you sign my noncompete?*

*Maybe.*

I stopped responding.

After digging around in my office for twenty minutes, I found the business card for Uri Bernstein, an attorney a friend had once used to get him out of a securities-related jam. Bernstein checked all the boxes as far as I was concerned: he was a Jew, and he was over fifty. When I asked his secretary for him, he took my call right away. I explained the situation and emailed him the noncompete. He glanced at it as we spoke and told me that for a thousand dollars, he'd draft a rescission of the noncompete for Bryce to sign. It was relatively boilerplate, so he could probably get to me by the end of the following day.

"Get it signed in front of a notary," he said, sounding very Jewy and wise indeed. "Otherwise, you and I could end up speaking again about the same subject."

"If this guy signs and reneges," I said, "I'll need a lawyer for a different reason."

"I can recommend someone if it comes to that, Mr. Fisher," Bernstein said.

He transferred me to his secretary, and I gave her a credit card number. By the end of that day, not the next, Bernstein emailed me the rescission. I texted Bryce.

*I can meet you with the paperwork tomorrow.*

*Email it to me, and I'll have my attorney look at it*, he said.

*The next complaint goes out Monday if it's not signed by then.*

*Get it to me tonight then.*

I spent most of the day in a celebratory mood, smoking weed and doing lines of coke and listening to music on my laptop. I couldn't remember the last time I felt so good.

I pictured Bryce in his big glass office reading and rereading the complaint, reading it on the can, reading it at his desk, his attorneys saying to him, "Level with me. If they investigate these claims, how many of them are going to prove to be true?"

Bryce was scrambling, I could feel it. I could almost see his red face as he paced his office floor, threw a phone, drank a shot of bourbon and then another. He called his wife and bitched her out about some unrelated matter. God, it felt good, like we were a pair of voodoo dolls, and I'd finally plucked the needle from my chest and driven it into his. I could breathe.

By one a.m., I was so fucked up and in such a good mood, I was standing out back in my robe with a cigarette in the corner of my mouth, taking measurements. I was going to rebuild that fucking deck come hell or high water. I'd spent about a decade in construction between all the other shit I'd done, and, like everything else I'd learned, it was still in there. I could build a deck with my eyes closed. I got online and ordered the lumber to be delivered the next day. As I finished, my phone pinged. I glanced at it expecting an order confirmation. It was Bryce.

*Yo.*

*You signed it yet?*

*No.*

*You'll feel better when you do.*

*Why?*

*Because you'll be slightly less of an asshole.*

*You got another job lined up?*

*How the fuck would I line up a job when I can't accept one?*

*Why do you want out of it so bad then?*

*Are you retarded?*

*Lol.*

*Sign it, and we never have to talk to one another again.*

I watched the dot dot dots appear and then disappear as Bryce debated his reply, drafting and redrafting it. Finally, it arrived. *You could come back to work at Keystone if you wanted.*

I laughed. Actually laughed. *I will never work for your dumb fucking ass ever again. Sign it or go to jail. At this point, I'll be happy with either outcome. I'm going to bed.*

The next time my phone pinged, it was Tonya.

*Fifteen minutes away. Feel like company?*

*Yessssss.*

By the time she got there, I was bouncing off the walls.

"You look totally fucking coked," she said.

"Why do you keep saying that?"

"Because every time I see you lately, you look completely crazy. Are you doing like way more than you used to?"

"Not way more. Maybe a little more. Listen, maybe don't walk in and start criticizing me right off the bat."

"It's not a criticism. It's an observation. I'm getting worried."

"I'm celebrating, honey. Is it that weird to get a little fucked up when I'm celebrating?"

"Well, tell me about it," she said. "I could use some good news."

I told her about selling the guns to Hector and about my brilliant plan to get out of my noncompete at last. I rolled a joint in the process and lit it.

"Haven't you had enough?" she said.

"What?"

"The joint. The wine. The coke. Like how much do you need?"

"I'm trying to have a good time. What were you expecting at two a.m.?"

"I want you to have a good time, Wince, but you're already totally lit. I kind of liked it when I could come over once in a while without you being this fucked up. I'm around fucked up people all day, basically."

"Well, that's not my fault."

"I know. It's just...maybe I need to join a running club or something. Like, I can't remember the last time I had a conversation with a sober person."

"I'm sober sometimes. But not at two in the morning. I mean, Jesus."

"Listen, I think I need to head out. I'm totally not trying to start anything. You're fine. I think I'm just having an off night. I should have gotten a room or something."

"A room? Why can't you go home?"

"Listen, I'm glad you're having a good night. Sorry to be a downer. I'll talk to you soon, okay?"

Tonya left. Whatever. I didn't let it get to me. I stayed up getting high and strategizing ways to reorganize the kitchen cabinets.

# 37

I WOKE UP late the next afternoon, took a leak, and went to the front porch for a smoke. I squinted out into the yard where, mysteriously, a stack of lumber had appeared. Oh, right. The deck. Last night's version of me calling the shots again.

My phone showed missed texts from my mother, my daughter, and a quiz from Luis about which Harry Potter character I would be, as if I had any fucking idea who any of them were. I saw I'd called Ruck sometime around three a.m. the previous evening. We'd spoken for seventeen minutes. I searched my mind but couldn't remember a single thing about the conversation. At least I hadn't called an ex.

I had two voicemails from Home Depot asking where on my property I wanted the lumber delivered, and one from Timothy asking how things had gone with the complaints. I flicked my cigarette into the yard and went inside to do a bump off the kitchen counter. I ordered my lunch and stood in the kitchen rereading the messages between me and Bryce. I wished there were more of them.

The person I most wanted to talk to was Todd. He knew what working with Bryce was like. He'd been there for some of the fun times and the end of the fun times. He answered on the first ring.

"You out?"

"What do you mean?"

"You out of the noncompete? I heard you were almost out."

"Who'd you hear that from?"

"Anna." Anna was Todd's old assistant, who still worked at Keystone. She'd never taken much of a shine to me, but she apparently still kept her ear to the ground. It wasn't worth asking how Anna would have heard. Bryce was notoriously bad at keeping secrets.

"Yeah, well, don't congratulate me yet. He hasn't signed."

"He will. And, seriously, I might have found a job for you."

"At Trinity? Praying over lunch? Man, I don't know."

"I'm telling you, they're making ridiculous money over there. At least call and meet with the guy. The thing I keep thinking is what if you and I both go there, figure out how they're doing it, and then start our own thing? Anna will come. And I know Timothy would. I talked to Amy about it, and she loves the idea."

"Amy hates me."

"Are you kidding? She's your biggest fan."

Bryce's next text arrived while I was at Luis's. Luis and I had returned to our old rhythm of hanging out primarily when I was buying coke. I was no longer expected to check in daily. As much as I appreciated his friendship, it was a relief not feeling so responsible for him.

*Tonight at my boathouse?* Bryce texted. *Eight o'clock?*

"That from your girlfriend?" Luis said.

"No, the noncompete asshole. He wants to meet."

"Alright, maybe tonight he signs."

"Maybe. Or maybe he finds some new way to fuck me over."

"Nah, you're too smart to be stupid twice."

Basically, I agreed. I was smarter than Bryce, it went without saying. My one great lapse in judgment had been the signing of the two-year agreement.

*I'll be there at eight-thirty.*

I went back to the crib to sober up and saw that the stack of lumber I'd ordered had become completely unstrapped and had slid down the slope of my lawn. I could picture the way it would look in another six months if I left it that way. The HOA already loved fining me for leaving my trash cans at the curb. They'd go into convulsions over this shit.

As I always did when I didn't know what else to do, I went to bed. I lay there for a while thinking about torturing Bryce. I felt amped up. The day had finally arrived when he might release me from the noncompete. I hated thinking of it that way, as him releasing me. Although at least I'd been the victor in the end. Or so it appeared.

# THE HAWK

I remember when my old friend Garland introduced me to a guy who was into falconry. Garland was thinking about getting into the sport, if that's what you'd call it, and he was trying to convince me how cool it was. Obviously, it seemed completely moronic to me. I couldn't figure out why anyone would hunt with a bird when you could hunt with a gun. A bird you have to feed and take care of and exercise and clean shit out of its cage.

I rode out to the middle of nowhere with Garland, and we met this guy in a field, standing there with a big red-tailed hawk tied to his arm. It had a white chest and these incredible brown and white wings. The fucking thing looked tough, I'll give it that.

It blew my mind seeing how the guy would set it free, and the hawk would sail off and take out a passing dove and bring it back. The thing that surprised me most was the way the bird kept returning to the guy. After a while, it actually kind of pissed me off. No hawk in its right mind wants to be tied to some dipshit's wrist, to get shoved into a cage and go from place to place in the back of a fucking Subaru when he could be flying around free all day. The whole time the guy was talking, I was looking at the little leather straps the guy held to keep the bird in check. By the end of the day, I was trying to mind control the hawk into escaping. What I wanted most was for it to take off towards the horizon and to see the expression on this guy's face when it did.

# 38

THOUGH I HADN'T intended on it, getting to Bryce's house over an hour late was actually kind of a nice final fuck you after all the shit he'd put me through. I passed his wife's car in the driveway, steering down a narrower drive that led to the boathouse. I'd been there probably three dozen times. The lights were on.

I went to the door and didn't knock. I'd never knocked before, and I wasn't going to start now. The place hadn't changed much. Same dark paneled walls. Same pinball machine. The only new addition was a framed poster of Anna Nicole Smith, back when she was still *Playboy* material and not just another fat gold digger. And there was Bryce, sitting beside a fire, drinking what had to be an old fashioned out of a lowball.

"Well, let yourself right in," he said. He was smiling his evil little grin. Fucking Bryce.

"What's up?" I said.

"Have a seat. I'll get you a drink."

As I sat, Bryce got up and went behind the bar to pour me a tequila.

"Good shit, right?" he said when I took my first sip.

"Yeah, it's not bad. You got your signing hand ready?"

"Listen, I want to talk about this. You weren't really going to try to report me to all those places, were you?"

"Are you kidding me?" I said. "I already reported you once. You know I'll keep going."

"So, you really sent that complaint? The one you sent me?"

"Yep."

"See, I still don't understand how you could do that after everything."

"You fuck up my livelihood, I fuck up yours. Seems like a pretty even trade."

Bryce gave me a look like he couldn't believe what I was saying. "My attorney drew something up for you to sign that says you'll cease sending complaints and withdraw whatever complaints you've made," he said.

"Why the fuck would I do that?"

"Because you want me to sign the noncompete release. This is fair. You get something, I get something."

"Gimme a sec." I took out my phone and texted Timothy. *That complaint I sent. Is it possible to withdraw it?*

As always, he replied instantly. *No. They either choose to investigate or they don't. There's no complaint withdraw process. Why would you want to withdraw it?*

*I don't.* I texted him my email password. *That's my password. Can you login to my email right this second and send the other complaint? It's in my drafts folder. All you have to do is hit send.*

*Right now? Why can't you do it?*

*I'm sitting here with Bryce. He wants me to withdraw the complaint.*

*Logging in.*

"Who are you texting?" Bryce said.

"It's personal."

Timothy texted me a second later. *You sure?*

*Yes.*

*Sent.*

"Show me what you want me to sign," I said.

Bryce went to the bar and came back with a sheet of paper. Like I knew it would be, it was some bullshit thing I was sure would never stand up in court if he was ever dumb enough to take it there.

"Here's a copy of the noncompete rescission you emailed me. We can sign at the same time."

Suddenly, the words of the attorney started to bug me.

"I want a notary," I said. "If we're doing all this signing bullshit, I want a notary here."

"You don't think we can trust each other?"

"I sure as fuck know I can't trust you," I said. "Whether or not you trust me is immaterial."

"Fine, we can go to a bank tomorrow. Or meet at the office. Alan is a notary. And Lisa."

"Which Lisa?"

"Lisa who you used to date."

"She moved to Dallas."

"You didn't know she was back?"

"No."

"Well, she is. I lured her back."

"I bet you did. I'm not waiting until tomorrow. We sign this shit tonight, or I'm going to keep sending out complaints."

"What notary is going to come out here at this time of night?"

"I don't know. But you should try to figure it out."

And so Bryce sat on his phone texting away and finally said, "Alright, found someone. They'll be here in half an hour."

"See, that wasn't so hard."

It's always like this when things are so heightened and everything is all fucked up: there's this weird pull towards normalcy. He poured me another drink, and his phone pinged on the table. He looked at it and said, "This fucking guy."

"Who?"

"This asshole I know. He's asking if I'm downtown. A hundred percent chance he's already trashed and doesn't want to get another DWI. I'm not going to respond."

"I wouldn't either," I said.

There was a pause, and then, for some reason, I asked how things had been at the office.

"You know it's actually getting better. Like, a lot better. People are so used to the online thing now. Even you could be good at it if you gave it a real shot."

"Nah, not my scene."

Bryce nodded. "You really could come back to Keystone, you know."

"You said that before," I said. "I assumed you were joking."

"I'm not joking. It could all be water over the dam."

I shook my head. "Too much water. You fucked me over."

"You fucked me over when you left."

"How is leaving fucking you over?"

"You left me with a lot of expenses."

"You could have left, too," I said. "You could have tightened your belt. I wasn't the one who wanted an office with glass walls and fucking wallpaper that costs three hundred a foot for some hallway no one goes down."

Bryce closed his eyes and took a deep breath.

"You know what would go good with this tequila?" I said.

"Yeah, I do."

He stood up and went behind the bar and came back with a CD case for *Kid Rock's Greatest Hits* with four fat lines of coke already laid out. They were much bigger than the lines I'd been doing lately. I guess this explained the late-night calls. He handed me the case and a straw. I bent down and inhaled.

By the time the notary knocked, the coke was gone. Bryce had done two lines. I had done three or four. It was speedier than the blow I'd gotten used to, and I felt my whole body vibrating and sweating.

Bryce took the CD case and put it behind the bar.

"Nose check," he said.

"Wipe the tip," I said.

"You too."

I was still wiping my nose when Lisa walked in the door. Classic Bryce. I hadn't seen her in months. I'd blocked her on everything. She looked fucking great, of course, in a short black skirt and a loose top that made her tits look incredible. Bryce put his hand on her back and leaned in to kiss her on the cheek. God, tell me they weren't fucking.

"Well, you guys look like shit," Lisa said.

Bryce looked at me and I at him. He shrugged.

Lisa carried a big handbag, and she reached in and took out her notary book, stamp, and ink pad. "Let's get this over with."

Whatever reaction the two of them were trying to get out of me, I wasn't giving it to them. I tapped my empty glass with my fingernail, lifted it over my head, and said, "One for the road?"

Bryce took my glass and refilled it. "I told you it was good," he said. "Lisa? Double gin and tonic?"

"Single. I'm not staying."

"What's your hurry?" Bryce said.

"I'm late to get fucked by an old friend, if you must know," she said, "and I'd rather be doing that than this bullshit with you two."

"Hardly the prim little office princess I've come to know and love," Bryce said.

"You want a princess, contact me during office hours. And don't make me deal with Winston."

"Oh, but I thought you two had parted on good terms," Bryce said, all innocent.

Lisa and I said nothing.

"Well, alright then. Let's get signing," Bryce said. "You need anything?"

"No, both of you just sign."

She handed a pen to Bryce and another to me. She took the glass from him. Lisa could always drink. She could drink, she could fuck, she could inhale coke like a fucking vacuum. The only thing she couldn't do was keep her legs closed.

I leaned down and signed, and Bryce did the same. Then we swapped and signed each other's documents. Lisa stamped them, initialed them, and recorded the transaction in her little book.

"Five hundred dollars," she said to Bryce. "That's what you said."

"You want it tonight?"

"I don't want it tonight. I want it right now."

"God, she really isn't all that fun after hours, is she?" Bryce said.

"No, turns out she's not," I said.

Bryce reached into his pocket and withdrew his wallet, from which he counted out five one-hundred-dollar bills.

"It doesn't cost anything to be polite," he said.

Lisa snatched the money from his hand.

"That's true," I said. "It doesn't cost anything to behave like a decent person. Not that either one of you would know."

"Wonder if a little coke might lighten the mood for you two," Bryce said.

As I knew she would, Lisa hesitated, then sat down beside him.

I stood and took the signed paper from the table.

"I'm good," I said. "See you later."

"Stay!" Bryce said. "At least for a drink."

I walked out the door without looking back.

Bryce followed me. "Hey," he said. "Wait."

I ignored him and kept walking, my feet crunching on his expensive black gravel.

Bryce stood in the doorway and shouted after me, "It was you who wanted the two-year noncompete, Wince. You were the one who was so sure everything was going to work out. You said, 'I'm not going anywhere. Hell, make mine for two.' You went around bragging about it to everyone. You claim to have this great memory, but you're—"

Who knows what he said next? I was in the Tahoe driving away.

One thing I've always noticed about these big occasions when you finally get your way, when you think you're going to feel that big sigh of relief: most of the time that feeling never comes. I'm sure I felt something, but whatever it was, it wasn't enough to uncoil the spring that had tightened in me all those months.

## 39

THREE DAYS PASSED, during which time I did not answer the phone. I avoided looking at it altogether. Of course, that didn't stop the flood. Luis texted, my mother texted, Sam texted, Hector texted. Even fucking Bryce texted to see if I wanted to go to dinner. I answered the ones from Sam, of course. And my mother, to keep her off my back. But mostly I spaced out and tried to shield myself from the world's constant intrusions.

On the fourth day, Hershey at Trinity called, as I expected he eventually would. He wanted to set up a time to talk, so I trimmed my beard and found a dress shirt that would look alright on a Zoom call.

I cleared away enough crap in my home office to make it look like I was a regular Joe tapping away at a keyboard in a blue Oxford. Hershey appeared on screen in a grey suit. He was about forty-eight years old and twenty-five pounds overweight, mostly bald with grey hair on the sides, and a trim beard he dyed dark brown. On the wall behind him a glittery, cursive decal read "blessed." Also visible were a framed flag folded into a triangle, a picture of his family, and a black-and-white "art photo" of a newborn baby curled up asleep on top of a football. In wealth management, if you're under thirty-five, your youth is your entire pitch. To anyone over fifty, you're basically an infant, fresh faced, charismatic, eager to build a little money for your clients so you can build a little for yourself. If you're over forty, there are more ways of being in money management. You can be a family man, a Christian, a patriot, or a jock. The best performers are a combination of two or more of these. Arnold Hershey at Trinity Wealth ticked all the boxes. He was a Christian first, a patriot second, a family man next, and an over-the-top Cowboys fan last.

"Congratulations, Mr. Fisher. I heard you finally freed yourself from a sticky noncompete."

It was one of those conversations where if someone asked you about it later, you wouldn't be able to recall a single word you said. You just knew you'd said the right things. By the end of it, I was hired, and so was Timothy. He wanted us to start right away, but I acted like I had a lot of personal stuff to wrap up and that Timothy would need a week to give notice.

I called Tim.

"Hey, boss."

"Hey. You cool with starting at Trinity in a week? I just got off the phone with Hershey. He wants both of us. I told him you're making seventy-five thousand, and he said he'd match it."

"Seems like sort of a stodgy office."

"Yeah, well, we'll be running the place in no time. Or figure out what they're doing and go start our own thing. That's what Todd has in mind."

"If you think it's the right move," Tim said, "I'll give notice today."

"Sounds good. We'll talk soon."

I went to the kitchen and did a bump, then another, then one more. Then I took what was left of the coke I had and put it in the safe in the closet.

I've always been able to quit coke when I wanted to. I've gone months without using anything other than weed when I needed to slow down. And I figured this was one of those times. I sat there thinking about how this could be a new start for me. Even if I was starting over at fifty-seven, maybe this would be the next thing I did to reinvent myself. I knew it wasn't the final answer, but why couldn't it be the thing that led to the thing that changed everything?

Hector texted.

*Hey man, how are they coming? Got the five ready?*

The five ARs. Fuck. I'd practically forgotten them. I didn't have the parts, and I hadn't come up with a plan.

I waited half an hour to respond.

*The five aren't ready, but I'll get them to you. Thing is, I got a new job and I need to focus on that for now. I'll do these five first though. By the end of the week.*

Hector started tapping out a reply, but I never got it, and I didn't hear from him again that evening.

# 40

QUITTING COKE IS always the same. You spend the whole day thinking about this tiny act you're not going to do for a while that changes your mood very slightly. It seems like it shouldn't be that big of a deal.[1] You remind yourself over and over that you're not doing it, and the whole thing makes you feel like kind of a loser, which makes you want to have a little coke, which resets the loop. You've already told yourself you don't really need it, that it's a habit not an addiction, all that stuff. Which is true, in my case. But if I'm honest, if I'm not doing coke, I basically want to sleep all the time.

That's pretty much all I did that week. Tonya asked about coming over. The first two times, I said no. The third time, I

---

[1] Years earlier, after the motorcycle accident, I had to take OxyContin for the pain. Those pills are fucking magical. They do exactly what they're designed to do. They make the pain go away. When I was on Oxy, I had a completely normal life. I sprang out of bed in the morning. I got stuff done. After getting back surgery, they kept me on Oxy. It was great. Then, I got sloppy and came up positive for coke on one of their mandatory drug tests, and suddenly I was cut off. The pain clinic wouldn't give it to me anymore. Everyone on the news and in all these movies acted like Oxy was this big, horrible drug that no one could get off of, but I never took it again, other than a few times when I bought some from a buddy or something. It wasn't that big of a deal. I didn't have it, so I didn't do it. It sucked that things went down like that, because I ended up having to get accustomed to pains I knew I shouldn't have to live with. I could never get over it. Whose business was it what drugs I took? Especially when I could quit them anytime I wanted.

didn't reply. I knew it would piss her off, but I couldn't deal with her or anyone else. It was bad enough I had to figure out how to get Hector his guns.

The usual flood of texts arrived, of course, with the additions of Timothy, who naturally wanted to know more about the job he'd agreed to take, and my ex-wife, who apparently thought we should start discussing our daughter's wedding, even though it was at least a year off.

The day the housekeepers were supposed to arrive, I texted and told the lead cleaner not to come because I wasn't feeling well. She texted back, *Please next time you tell me sooner so I can book another client.*

When I was awake, I thought about the two things I absolutely had to do: build the five guns for Hector and find a suit that fit. The question that kept me stuck was whether I was really going to do either of them without coke.

Then it was Thursday. I was supposed to start work on Monday. Hector had sent one of his pushy texts the day before, or maybe the day before that. The problem was finding parts. I had enough auto-sears on hand, but for the rest my only hope was to start driving to gun stores, praying they'd have what I needed. Somehow, I got up at 10:30, showered, and got out the door by noon.

Gun stores are always a drag. Which is weird because you'd think they'd be cool. It's the same problem as shooting ranges. Bunch of guys with chips on their shoulders, all these little rules and prohibitions, acting like they're selling grenade launchers when really they're dicking around, receiving orders, and filling out paperwork.

For once, I was the ultimate chump. I overpaid at every spot in town. Inventory was shit, what with everyone thinking the world was ending. I was so desperate, I even paid some clerk at McBride's $350 to sell me one of his own AR-15 lowers with an M-16 pocket. My profit margin was disappearing, but what choice did I have? After scoring four bolt carrier groups at Red's, I went home and slept like I'd run a marathon. Finally, I had everything I'd need for five full-auto ARs.

After a nap, I DoorDashed Chinese and texted Hector that I'd meet him at Luis's on Saturday with the guns. Of course I didn't build them on Friday. I waited until the last minute.

# 41

SITTING ON LUIS'S couch, passing the joint back and forth and waiting for Hector to show, I let myself imagine this might really be the end of my life of crime. Maybe I'd gotten it out of my system, and I'd go on to have a relatively normal life after all. I still hadn't done any coke. When Luis asked about it, I told him I was taking a break. But I was on the verge of folding by the time Hector showed up.

He walked in the door, looked at the two of us on the couch and said, "Hey, it's The Cheech and Chong Show."

"What up, Heck?" Luis said.

"Busy night. How's it going, Wince?"

"What's up, man?"

"Wanna go to the garage?"

"Sure."

I followed him out, and we opened the garage door. I loaded the guns into his Bronco.

"Six, right?"

"Five," I said.

I looked at him to see if he was fucking with me, and he was. He smiled and handed me a fat envelope with $40,000 in it. Four stacks of $10,000.

"So, you're done then? No more Mr. Machine Gun?"

"We'll see. I've got to start this new thing, see how it goes. I'm at least going to take some time to get adjusted."

"Crime's more fun," he said.

"I don't know, man. It's all fucking work. But I'll be around. You need something, you know how to reach me."

Then Hector did something kind of funny. He opened up his arms, took a step forward, and embraced me. It was pretty much a bro hug; nothing *Brokeback Mountain* about it as far as I could tell. Still, it seemed kind of out of character for him, and I wondered if he was so paranoid, he was checking to see if I was wearing a wire. It didn't offend me if that's what it was, and it didn't offend me if it was friendship. Hell, it was probably a combination of the two, which didn't offend me either.

When we went back inside, instead of taking off like he usually did, Hector sat with his dad and me and, for once, even puffed on the joint his dad had rolled. The three of us watched the last half of *Terminator 2*, and I left right after Arnold got lowered into molten metal.

I was still stoned when I got home, watching the ceiling fan from my bed, feeling stressed about the job I'd be starting in two days. I'd swallowed two Xanax to mellow out, and they were finally kicking in when I realized I'd completely forgotten about testing those last five ARs.

When I woke up on Sunday afternoon, I stood on a carpet of dirty clothes in my walk-in and counted all the suits I owned. Thirty-seven. Not one of them fit. I told myself it was only a matter of time before I lost enough weight to wear every last one of those suits again. Why spend a bunch of money on a body I wasn't going to have much longer? I wouldn't be this fat forever. I couldn't be. I looked it up and found that the Kohl's near my house closed at eight o'clock. I left my house at 7:30 and, for $350, I bought two polyester suits that felt like shit and fit like shit. When I put them on, I looked like shit.

# FIGHTING BOBCATS

When I was a kid, there was a bounty on bobcats. They used to pay two dollars apiece for their ears. As soon as my grandfather heard about it, he started spending every spare moment hunting them. He wasn't interested in population control or sport. He did it for the money. I went out with him all the time. I was five years old running out to these big dead cats he'd shot, sawing off their ears with his knife and bringing them back to my grandfather like crisp dollar bills. They were always dead when I cut their ears off, but I often imagined one day I'd go over to a bobcat to discover it had only been stunned, and the minute it felt the knife against its ear it would awaken and spring to defend itself.

The dream I dreamed most as a child was about a bobcat snapping to attention as I dug my blade in. I knew I'd have to fight it, and I wasn't afraid. In my fantasies, my victory was a foregone conclusion. The day it happened would be the day my grandfather saw me as a hero the same way I saw him as one. I might never be as wily and cunning and wise as him, but I could be courageous and strong. I told myself that one day he would be there to witness my brawl against a wildcat, and I would amaze him.

That fight never came to pass, of course. Within a few months, enough bobcats had been killed that the county no longer offered the bounty, and we stopped hunting them. Years later, there was a bounty on coyotes, and my grandfather was back at it, but I was in school by then, so he usually went out without me.

# 42

TRINITY WEALTH MANAGEMENT operated out of a three-story brick building in a part of town overwhelmingly occupied by medical and dental offices. Practical, but definitely not cool. I showed up early, believe it or not. Dressed in my shitty suit, I sat in the parking lot eating an Egg McMuffin at seven thirty. I hadn't had any coke in a week and a half, but I sure could have used it then.

Finally, I went inside, took the elevator to the third floor, and found their office. Arnold Hershey was there, dressed in a three-piece suit that made mine look like Armani.

"Look who's here," he said when I walked through the door. "Our newest team member."

I followed him into a conference room that couldn't begin to rival the one Bryce had paid a fortune to appoint so outrageously at Keystone. Timothy was there, looking every bit the part of the preppy schoolboy. We sort of rolled our eyes at one another over the formality of the event, being introduced around a room full of people, almost all of whom were related in some way to Arnold Hershey. You would have thought the guy was a Mormon.

Things started going sideways when Hershey's wife showed us where Timothy would be working, a desk in a swastika-shaped quartet of cubicles.

"Oh, we always work together," I told her.

"At Trinity, the processors all work in this area."

Timothy saw the look on my face and said, "It's okay. This is fine."

"Yeah," I said. "We'll figure this out. This will be temporary. Even if we end up just putting an extra desk in my office."

Hershey's wife already looked burnt out by us.

She introduced Timothy to the dweeb at a neighboring desk who she said would be training him. I gave Timothy a quick look to suggest that this was but a momentary blip. And although I was worried, I wasn't worried enough.

We left Timothy behind for the remainder of the office tour and eventually arrived at my desk. Not my office. My desk. In a fucking cubicle. I don't know what my face suggested, but I know I didn't do a great job of hiding my disappointment, something Hershey's wife clocked with pleasure. I thanked her, saying her name, as one does. She gave me a look of horror, at which time I realized I had called her the wrong name but I couldn't recall her actual name, so I let it go.

"You can get settled until ten, then meet Arnold in his office," she said, and then she stomped off in her ugly shoes.

Describing the experience of that week is boring even to me, so I'll skip to the punchline. I shadowed Hershey for a week, watching him in a series of lame Zoom meetings with other religious fanatics, who were his only clients. Timothy and I barely saw one another, went out for drinks twice, and agreed we'd stick it out as long as possible. Then, on Friday, after a lengthy argument with his wife about the way I preferred my tax documents to be handled (I wanted the write-off perks of a statutory employee), Arnold called me into his office. I sat down, and he said, "This isn't working out."

"No discussion? You're just firing me?"

"Listen, Winston. We both know you're great at this job. But I can't have anyone working here who can't get along with my wife. In my experience, it's better to cut things off at the start when you know it's not going to work out. I respect you, and I thank you for your time. I'm sure we'll see one another again."

He rose to shake my hand.

I had lasted a single week. And though I hated the job, I felt like sobbing. I felt like crumbling to the floor and holding my knees to my chest and screaming. I felt like gripping Arnold by the back of his head and slamming it into his desk until his teeth were shattered and blood gushed from his mouth.

I thanked him as graciously as I could for the opportunity and left his office.

I went to my desk, collected my things, then went to Timothy's desk and asked him to come talk to me for a minute.

"They fired me," I told him when we were outside, a lit cigarette already between my lips.

Tim looked at me, registering what I'd said.

"Jesus, you're kidding. Because of the tax thing?"

"I guess 'don't fuck the boss's wife' has more than one meaning."

"Well, should I—"

"I'm the only one who got fired. Go back inside and keep working. Hey, at the very least, you got a raise, right?"

"I guess," he said. "Man, this fucking sucks."

"Yeah, don't make a big deal of it. He's right, it probably wasn't a good fit. We'll figure it out. You're doing great here. Stay the course."

Timothy gave me a hug and went back inside.

I almost made it home before I started to cry.

During the eighteen months of my noncompete, I sometimes wondered if the problem was that I had no one to impress anymore. No bobcats to fight for my grandfather, no wife or girlfriend to keep from complaining. I wasn't picking up tabs every time I got a nice bonus or trying to break every sales record. If I had still been married or living with someone, I honestly don't think I would have wasted my days so carelessly. Even when I was with women I didn't like anymore, I didn't want to give them the pleasure of seeing me at my worst. I guess I was hopeful that, at Trinity, under observation, I would change. Now, that theory had been totally disproven.

By the time it got dark, I was on two Xanax, four 25mg weed gummies, two bottles of wine, maybe twenty bumps of coke, and ten bong rips. I texted Tonya, and she said she'd come over.

She let herself in as she usually did, but I was half-catatonic on the couch. She must have said my name. I opened my eyes to see her standing over me.

"If this is what you're like when you're working, maybe you're better off unemployed," she said.

"This isn't me working. This is me fired."

"Oh, shit. Not really?"

She sat down beside me and took my hand.

"It's true," I said. "One week. I got into it with the boss's wife. Some stupid fucking thing. Fucking dyke hated me from the minute I walked in the door."

"Oh, honey. I'm really sorry. What can I do?"

"Nothing. Just wanted to see you tonight. Thanks for coming over. Sorry I'm so fucked up. Don't say anything about it, okay? I know I've been blowing you off a little lately. Believe it or not, I was clean for over a week before tonight."

"You don't have to explain anything. I'm sorry things went sideways with your new gig."

We went to the bedroom and got undressed. I knew she didn't have any illusions that we might be able to have sex, but for a time she did lay on top of me and wrap her arms around me.

"You always land on your feet," she said. "I'm not worried."

"Wonder what I should tell my mother and Sam."

"Tell them you quit."

"That's almost as bad as getting fired to them."

When I woke up to piss a few hours later, she was gone.

I went and stood on the back deck, having a cigarette. I was standing there looking at the moon when I heard a loud crack and found myself on my ass. I didn't know what had happened. All I knew was that my leg hurt. Then I snapped to reality and saw why. My fucking foot had gone straight through one of the rotting boards. One leg was dangling clear through to the other side, and the other was straight out in front of me. The perfect ending to the perfect fucking day. At least I hadn't landed on my balls.

I had to summon all my strength and pull off some seriously superhuman maneuvers to free myself. No matter what happened, I couldn't be found that way, naked except for a robe, spread open like a raw chicken. What kept me going was the fear that I'd have to cry out until one of my neighbors heard. Which one of them would come to my rescue—the potbellied Indian to the left or the commie to the right? By the time I got free, I was covered in sweat and breathing so hard I thought I might have a stroke. My leg was all scratched up, my ankle was bleeding, and my robe was torn.

Panting, I smoked another cigarette braced against the house. I called Ruck and went to voicemail four times before he finally picked up.

"Fired, huh? Well, don't take it personal. Same thing happened to Oprah Winfrey, you know. She got canned, and it changed her whole life, so they say."

"The fuck kind of a thing is that to say to me, Ruck? Are you fucking kidding me?"

"Yeah, calm your tits. I'm fucking with you."

"I'm so fucked, man. You have no idea how fucked I am. You know Sam's getting married?"

"Still can't believe she's old enough."

"You got any idea how much money a wedding costs now? I needed that fucking gig. That bitch. I could kill her. She had it out for me the minute I walked in there. This is probably some weird shit they do over and over again. They hire some innocent guy, she doesn't like him, then she gets to use her influence to get her dimwit husband to can him. Then she gives him a blowjob or something. Some weird game like that, you know?"

"Yeah, well it sounded like a shit job anyway."

"To say the least."

Ruck exhaled a lungful of smoke. "Well, there you go."

"Wonder if I could sue."

"Wince. It's the middle of the night. We both know what you're going to do. Let's just talk it through."

I hesitated to use the phone for stuff like that. It was a bad idea. But I'd had plenty of bad ideas, and it had never stopped me before.

## 43

I TOLD MY mother and Sam that Hershey at Trinity had tried to get me to sign another two-year noncompete. When he told me he wouldn't keep me on without it, I told him to eat shit. Mom was outraged, of course. I listened to her rant against Hershey, relishing the brief window before she returned to being on my case about what I'd do next. She wondered aloud what she should tell her friends, who had all been so relieved to hear I'd finally gotten a job. "Oh, they'll be so disappointed," she said.

Todd called when he heard about me getting canned, which meant everyone else in town had heard too.

Then came the text from Bryce. *Trinity is a shithole anyway. Everyone knows Arnold Hershey's wife keeps his balls in a jar.*

I didn't reply.

Hector must have heard it from Luis.

*Sorry to hear about the new job*, Hector texted. *Does that mean we're back in business?*

*Maybe. How'd that last load work out for you?*

*They're long gone, and I could use more.*

*I'm going to take a week or two off, but if you want another lift to the junkyard, I could help out. Wouldn't mind earning a little product now that I'm unemployed again.*

*Perfect timing. Wednesday if that works for you. Same time as before.*

It was six days away. I told him it wouldn't be a problem.

Tonya came over that evening without calling. I was sitting on the front porch smoking and looking at Facebook when she pulled up. She was holding a bottle of red wine.

"You're up and about!" she said. "How are you?"

"I'm alright. Been thinking up lots of nice ways to torture Hershey to death."

"Well, you look good. Look what I got you."

She held up the bottle. Caymus Cabernet Sauvignon. I couldn't remember telling her it was my favorite, but I must have mentioned it.

"And all I had to do was get fired."

We DoorDashed Wienerschnitzel. Their chilidogs reminded me of being a kid. Tonya had never had them, which I couldn't believe. I was always exposing her to things she'd missed out on.

After dinner, we watched the first half of *Enemy of the State*, where everybody's out to get Will Smith. We never paused movies to talk. We just talked over them.

I said, "Hey, there's something you might could help me with this week."

"Yeah?"

"It's something you couldn't tell anybody about. Not that you tell anybody about this stuff, anyway. You don't, do you?"

"I mean, I have friends who would be familiar with your name."

"Really? Like you talk about me with them?"

"Yeah. Is that not allowed or something?"

"No, it's fine. I just never thought about it. Well, I'd need for you to not discuss this with them, okay?"

"Wait. Pause the movie. What are you talking about?"

"This sounds weirder than it is, I promise. I need you to be available Wednesday night. I don't mean dropping by after everything else. I mean, you'd have to be here all night. I'll pay for your time, because I know it could cut into your other stuff."

"This doesn't sound like it's for sex."

"No."

"Well, what is it then?"

"I'm not trying to be all mysterious or anything, but I think it would be better if we didn't discuss it until after. It's nothing dangerous. You don't have to do anything but hang out here."

"With you?"

"No, I'll be out."

"I know you said it's not dangerous, but this sounds kind of dangerous."

"It's not. Definitely not for you. I just need to know if you can make it on Wednesday. Rain or shine, okay?"

"Yeah, I can do that."

"Alright then. Listen, I don't want to go over it all tonight. Let's watch the movie. I just wanted to be sure it was on your schedule."

Amazingly, Tonya let it drop. I guess she knew she had the whole week to pester me about it.

Before she left, she looked around and said, “While you’re out doing whatever you’re doing, you mind if I clean up around here a little? It’s getting more and more… chaotic, you know?”

She was right. That was the word for it.

# 44

I WENT BACK to one of the spots I'd discovered on my recent tour of the local gun shops. It was a new store I'd happened upon by accident. The first time I went there, they hadn't had anything I was looking for except charging handles, but something I'd noticed had stuck in my head: an Aero Precision 10" upper receiver in 300 Blackout.[1]

I bullshitted with the shop owner a while. He knew his shit, and he could see I knew mine too. Not that it stopped him from charging me nearly $800 for an upper I could have gotten online for $300. While I was there, I grabbed three boxes of subsonic ammo along with some paper targets printed to look like "bad guys," by which I mean a white dude in a black wool cap in a

---

[1] The problem with an AR-15 shooting the usual 5.56 ammo is that whether it's converted to full-auto or not, it's loud as fuck. You can put a silencer on it and dampen the sound so that you can hear it from a hundred yards away instead of a mile away, but that's it. And in a situation like mine, sound was the thing that would get you killed. Which is why I thought of the 300 Blackout. A complete upper assembly for a Blackout rifle fits right onto your standard AR-15 lower. All I'd have to do was pop it on and pin it into place. See, unlike 5.56, 300 Blackout can be suppressed. Like truly suppressed. A silenced Blackout with subsonic ammo sounds like a fucking pellet gun. Which meant maybe I could get some damage done before anyone pricked up their ears enough to come do damage to me.

variety of poses. I paid in cash, not that it would matter with all the surveillance cameras.

At home, I drilled out a couple of solvent traps and dug around in the gun room for an AR-15 lower I liked. There were at least ten of them I hadn't considered selling to Hector in case I needed them. Now I was glad I'd kept them. I popped the 300 Blackout onto the lower. I screwed the silencer onto the barrel. It was that easy. I held it in my hands. It felt solid and reasonable. Downstairs, I set it on my grandfather's table and turned on the TV. I watched *Man on Fire* and thought about the junkyard and didn't think about it.

When it got late, I went to have a smoke, turned off the back deck lights, and took the gun with me. I could hear coyotes beyond the fence. I aimed at a tree and fired. The sound of the bullet hitting the bark was louder than the sound of the gunfire. Hell, the sound of the gun's action was louder. I'd always thought of solvent trap silencers as pretty shitty, but I'd only used them on pistols, and mostly with supersonic ammo. This was a different thing entirely. It was so quiet, it was fun. In videos online, when you see guys firing silenced weapons—truly suppressed guns with great silencers—they always laugh. Every time, you see these guys burst into laughter. Now I got it. I fired at another tree, and another, and I laughed.

When I was sure his wife would be asleep, I called Ruck, who answered on the first ring.

"You're awake."

"Figured you'd call. You get the upper?"

"Yeah. Already fired it off my back deck. Very quiet. Fun to shoot."

"And Tonya?"

"She's all set for Wednesday."

"Great. What's next?"

"I need to zero the rifle."

"What distance you sighting it in at?"

"I'm thinking fifty yards."

"Maybe think more like twenty-five. That's what I'd do."

"Yeah, maybe."

"When's the last time you went shooting? Like target stuff?"

"Long time."

"Maybe something to think about. Get to the range if you can."

"Yeah, probably a good idea."

"I gotta go to bed soon, Wince. We had a birthday party for my step-granddaughter today. Seven-year-olds go buck wild."

"I remember."

"Should we talk about the ethics of any of this stuff?"

"The ethics?"

"The morals."

"No, man," I said. "I don't need to talk about that."

"Well, maybe another time."

When we hung up, I thought about the question he'd asked. It sort of bugged me that he brought it up at all, knowing Ruck like I did. The guys at the junkyard were criminals. I didn't mean it as a dig against them. I thought of myself as a criminal too. Being a criminal exposed you to being bettered by other guys who weren't playing by the rules. I never liked the idea of hurting an innocent person, obviously. But they weren't innocent. They were cartel guys.

The next day, I was up by ten. Converting the Blackout to full auto was a cinch after all the practice I'd had lately. I drilled the hole, ground off the serial numbers, pinned it all back together, and texted Hector to ask if I could go out to his land to test something.

*Anytime, man. Is it something you might sell me?*

*Nah, just tinkering right now.*

*Cool. Have fun. We're still on for Wednesday, right?*

*Of course.*

*You got that headlight fixed?*

*Yep.*

Of course, I hadn't replaced the headlight and had worried about getting pulled over since the first junkyard run. I took it as a sign of my ever-improving mental health when I stopped by an Auto Zone on my way to Hector's property instead of putting it off to the last minute. For twenty bucks, plus the cost of the bulb, a skinny Black kid behind the counter replaced it for me. Another thing marked off the list. It was starting to feel like a habit.

At Hector's property, I stapled targets up and did some shooting. I got the new Blackout sighted in at twenty-five yards like Ruck suggested and basically ran some drills. It wasn't great, but it went better than I'd worried it would. I'd always been a decent, though not phenomenal, marksman, and it appeared I hadn't lost it. I also brought two Glocks that I'd carry as backups. I'd shot them more than anything, aside from hunting rifles, but I dicked around with them for a while, nailing the targets in the head pretty consistently. All in all, I felt pretty encouraged.

Tonya texted when I was on my way back to the house.

*Still planning to be there Wednesday.*

*Sounds great. Thanks, honey.*

I was starting to think everything was going to be fine. Everything was coming together. Money was attracted to me. Anytime I reached into my pockets, I'd find money. I'd look into the safe and see money. In four days, everything would work in my favor. Positive thinking.

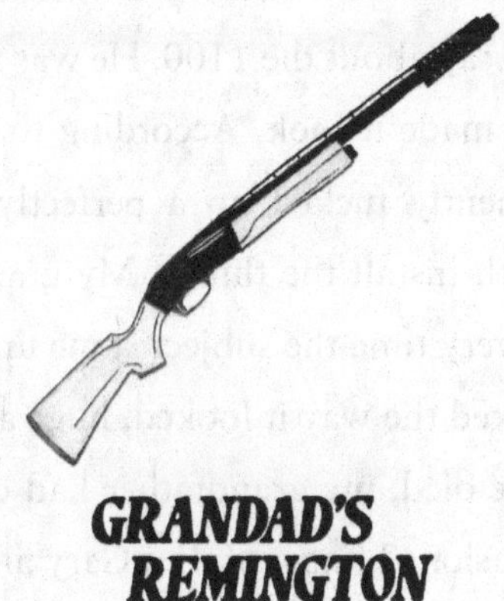

# GRANDAD'S REMINGTON

I REMEMBER WHEN I was forty-two. My grandfather had died just a few months earlier, and I took his Remington 1100 to be reconditioned. He'd lent it to me years earlier with no particular expectation of receiving it back. That's the way guns were treated in my family then. My father and I lent guns back and forth the same way. They were really our guns collectively. My grandfather's Remington was a beautiful shotgun, a 12-gauge with a Cutts Compensator on the barrel.[1] When he died, it was mine because it was in my possession.

My grandfather and I loved his Remington, which, for some reason, made my uncle Gary hate it. I never liked Gary, so his distaste for the gun made me like it all the more. Gary always had something to say about the 1100. He was fixated on the way the compensator made it look. According to him, my grandfather had permanently fucked up a perfectly nice shotgun by having a gunsmith install the thing.[2] My grandfather defended the Remington every time the subject came up, and I, of course, took his side. I liked the way it looked, huge and intimidating.

By the time he died, my grandfather had disowned Gary on at least three occasions I knew of, but Gary always found a way to weasel back into the fold. I always hoped that, in the end, Gary would find himself disinherited, but my grandfather died during an upswing in their cycle, and he ended up getting about fifty grand and a bunch of my grandfather's tools.

---

[1] A compensator attaches to the end of the barrel of a gun. Imagine a four-inch metal tube with a series of slats along the top and bottom. When you fire a weapon without a compensator attached, the gasses knock the gun back creating recoil, which causes the barrel to rise. Which causes the marksman to lose his target in the sights and prevents him from quickly recovering to fire again. When attached to a firearm, the slats in a compensator give the gas somewhere else to go. They go out the slats, which keeps the barrel from rising as drastically. Compensators are pretty common in handguns these days, but they're less common in shotguns now. They enjoyed a period of popularity in the 50s and 60s, largely because of a widespread advertising campaign by Lyman, the company who made most of them. One thing you can say for them is that they don't make a gun any quieter. In fact, they amplify the sound. My grandfather's 1100 sounded like a fucking cannon when you fired it.

[2] There were different ways of installing them, but the way my grandfather had it done was he had the gunsmith cut off six inches of the barrel and then add the compensator. Other people just tacked it on to the end, but it made the gun long as fuck and awkward to carry.

Shotgun parts take a beating, and they need to be reconditioned from time to time—bluing, springs, grips, anything that might not be functioning properly. That's why, after he died, I took the 1100 to a gunsmith. I felt guilty for not spending more time with him in his final years and was lonely for his company. I had the gun fixed up in his honor, in a way.

I had just pulled into my driveway after picking it up from the gunsmith when my uncle appeared in front of my house and said he wanted to borrow the old Remington for a weekend turkey hunt. I stood there stunned.

"Come on," he said. "Go inside and grab it for me, would you?"

"I just paid a fortune to have it restored."

"Did you get it restored for no one to shoot it, or for it to be shot?"

For a fuckup like Gary, the timing was incredible. He stood on my lawn, kind of smirking and puffing out his chest, looking like King Shit in a pair of coveralls. He'd never been to my house before. I wasn't even sure how he'd found it.

What could I say? Not lending him the gun would have created a massive family rift, which would inevitably lead to a flurry of phone calls from my mother. So I did it. I gave him the gun, only after he promised to return it the next week on his way back through town.

Of course, I didn't hear from him. My calls went to voicemail. I asked my mother about him, and she said she heard he was back in Houston. I called again and again.

Finally, one day he answered and said, "What is it, Winston?"

"The Remington. You were supposed to return it to me weeks ago. I need it."

"Well, you'll have to use something else. This is my gun, and it always has been. It belonged to my father. My father, not yours."

Then he hung up. A complete and absolute piece of shit. I could have ripped his fucking heart out. I didn't, though.

I went about it like a civilized person. For months, with my mother acting as intermediary, I did everything I could think of to get the gun back. I bought the exact same gun twice. One with the Cutts Compensator installed, one without. Gary wouldn't take the trade. I offered him both. Still no. I offered him cash. Then three of my grandfather's other guns that I still had. Ones Gary actually liked. The Remington was mine, and I wanted it. I couldn't have my grandfather back, but I could have that gun. Except I couldn't.

# 45

MY MOTHER CALLED. I let her go to voicemail. I knew it was a long shot, but I had this delusion that I could somehow delay speaking to her until everything was done. I figured I should do my best to stay focused and clearheaded.

A moment later, she texted. *Call me. Now. It's important.*

Two minutes later, she texted again. *Medical crisis.*

I called and she answered right away.

"What happened?" I said. "Is Dad okay?"

"What do you mean? Your father's fine."

"Who's having a medical crisis?"

"Your uncle Gary. They think he's had a stroke."

"What uncle?" I said. "I don't have an uncle Gary."

"Winston! Stop it. We're here packing up in a frenzy. We're going to try to make it to Houston tonight."

Oh, Jesus. My parents were coming to Texas.

"Now, don't go off half-cocked," I said. "What happened?"

"Jody found him in a heap at the bottom of the steps outside his house."

I stifled a laugh. "He's probably found Gary like that a hundred times. He's a drunk."

"He's your father's brother," my mother said, lowering her voice, "who happens to have an alcohol problem."

"He really had a stroke, or they think he had one?"

"That's what they're saying. I'm so worried about Jody."

Jody was Gary's bastard son whose mother was a Mexican housekeeper my grandfather used to pay to clean Gary's trailer when they were getting along. Jody was the reason my grandfather disowned Gary the second time. Jody was probably in his thirties by now, a true loser who hadn't managed to achieve anything so grand as cleaning his own fingernails last I'd heard. My mother kept me updated from time to time. He'd recently been fired from a cashier job at a gas station for stealing from the till. That was after he'd been fired from a carwash for grabbing some teenage girl's ass. Instantly, my mind flashed to my grandfather's Remington 1100. Knowing Jody, it would end up in a pawn shop about five minutes after Gary died.

"Poor Jody? Are you kidding?"

"Well, he didn't grow up in a stable household like you."

I could've commented on the stability of the household in which I was raised, but I didn't.

"If he wasn't related to us, you wouldn't be caught dead within a hundred yards of an asshole like Jody."

"Oh, this is him calling," my mother said. "I'll call you back."

I thought about my grandfather's shotgun, a gorgeous weapon with checkering on its stock and curlicue engraving on its black receiver. They didn't make guns like that anymore, American walnut with a high gloss finish. Even though I had one exactly like the one my uncle stole from me, it wasn't the same as having my grandfather's. Like it wouldn't be the same going out and buying a replica of the '64 Mustang I drove as a teenager. It would look the same without having the weight of history.

I hadn't thought about that gun in months. It was like someone put on an old record I hadn't heard in a long time, the way my mind could turn it over. That fucking shotgun, stolen from me. Gary, that worthless piece of shit. Jody, that greasy slimeball. The shotgun was mine. I'd paid hundreds to have it fixed.

My mother called back and told me the doctors had confirmed it. Gary had had a stroke and was laid up in the hospital with half his face all fucked up, though according to Jody things might still get back to normal in the days ahead. The doctors were optimistic.

"He doesn't want us coming," she said. "Can you believe that? Gary's in bed, half paralyzed, and he somehow manages to tell Jody he doesn't want us coming there right now. He said maybe next week. Don't you think your father will be upset we're not going?"

"No," I said. "I think he'll be relieved. I would be."

Twenty minutes later, my mother was still talking when she suddenly realized she should probably go downstairs and tell my father to stop packing the car. The poor bastard.

There's a habit I have, when I'm in the mindset of getting things done, which is that I imagine I should take on more. It's only a three-hour drive, I thought. Maybe it would be a good warm-up for the junkyard.

The route to Houston materialized in my mind, a blue line tracing a map.

It was still Saturday. It was still light out. The junkyard trip wasn't until Wednesday.

I sat and thought about how I'd spend the days ahead. I couldn't see myself going out to Hector's land every day firing at paper targets. I'd probably sleep the days away, losing my edge, like a flag going limp when the wind dies down.

"If you have the balls," I told myself, "you have the time."

# 46

IN THE GUN room, I found the two Remingtons I'd bought back when I was trying to entice Gary to give me back my grandfather's. The one with the Cutts Compensator was the one I was looking for. I dug around until I found a soft-sided shotgun bag and packed it inside. Downstairs, I looked up Gary's address in the tax records and plugged it into my phone. It was a two-hour and forty-seven-minute drive. If I left right that minute, I could beat traffic out of town. I stuffed a change of underwear, socks, and an extra T-shirt into a backpack, dropped in a Glock, and left.

I had to stop once for gas and again at Whataburger, but my brain was switched off. I didn't look at the things I was passing. I didn't think about what I was going to do. I sort of looked at it out of the corner of my eye then turned my head to forget about it. If you did things that way, when it came time to act, the deeper part of your brain already knew what to do. That was the theory, anyway.

Gary didn't really live in Houston, it turned out. He barely lived in Harris County. My GPS guided me out to a brushy shithole where you could smell the gas refineries. It was dark by the time I stopped for cigarettes at a little family-run filling station with ancient gas pumps and a gum-chewing hippo behind the counter. I was close enough to Gary's house I figured this was probably where he bought his daily sixer of Natural Light.

I was in and out as quick as I could be. It was dumb of me to stop there. Maybe I was trying to slow things down a little. Every mile had me smoking more, feeling edgier. Asking myself what the fuck I was doing and then having to tell that part of me to fuck off.

I got in the Tahoe and took a breath. I wanted to sit there and smoke a cigarette, but I thought if I did maybe Jody would pull up and see me. But then, maybe if I went to Gary's right away, Jody would find me there. I did a bump off my knuckle and called my mother.

"This day!" she said. "I swear."

"What's happening? How's Dad?"

"He's alright. He's worried about his brother, of course."

"Right, I forgot. What's happening there? He dead yet?"

"That's not funny, Winston. I just got off the phone with Jody."

"How did he sound?"

"Tired. He's been at the hospital all day."

"Jody doesn't actually live with Gary now, does he?"

"Well, he moved out for a little while, you know, with a friend of his, but I think that didn't work out for some reason."

"So, he's still at the hospital now?"

"Yeah, he was going to leave soon. You know how it is. I told him, 'Visiting is almost as exhausting as being the one stuck there.'"

"Which hospital did they take him to?"

While my mother talked, I looked up the hospital. It was twenty minutes away.

By the time I was able to get her off the phone, I was parked off a farm-to-market road in front of an isolated single-wide that was apparently the shitty abode of my uncle Gary. There wasn't even a porch out front, just a set of steps. I almost smiled. It was exactly the place Gary deserved to live.

I turned off the engine and got out. I was sweating.

Don't be a pussy. This is nothing, I told myself. It's nothing compared to what you have to do on Wednesday.

At the bottom of the steps was a patch of mud where Gary must have landed.

It was dark out there, hardly any light except for my headlights. You could actually see some stars. It was quiet, too.

I got the shotgun case from the Tahoe and went to the front of the house, avoiding the mud as much as possible. I used my shirt to try the knob. It was locked, as I assumed it would be. The windows were too high to go around checking them. Fuck it, I kicked the door. Once, then a second time. I was preparing for a third, already gasping for air, when a key balanced on a light fixture next to the door fell at my feet. I tried it, and it worked.

I leaned against the frame a moment thinking maybe it was true that when you take action, any action at all, God or the universe or whatever rises to meet you halfway.

The moment I turned the knob and opened the door, I realized I'd forgotten the Glock in the Tahoe. At the same instant, I almost found myself in the mud like Gary, because a little black and white Chihuahua about the size of a football dashed out the door and down the steps, scaring the shit out of me in the process. I turned around in time to see it run past the Tahoe and in the direction of the road. Then it was gone into the darkness.

Shit. Jody could be back at the trailer at any minute. The thought of running after the dog made my heart pound. My second wife, the whore, described my gait as "lumbering," which was both accurate and unnecessarily unkind, her specialty.

Worry about the gun first, I told myself.

I stepped into the dark trailer and was immediately struck by a wall of stench. Stale cigarette smoke, dog shit, and unwashed dishes. It was overpowering. Maybe I'd done the dog a favor getting him out of that fucking trailer. But I knew that was only partially true. Odds were good he'd be snatched up by a coyote before long.

In the dim light, I could see one of those tall floor lamps with a flared top. I debated it, but, hell, the Tahoe was already outside with the headlights pointed at the house. I switched it on.

Single-wide trailers are themselves a kind of hell, as far as I'm concerned. If anyone were to challenge me on it, I'd be the first to admit that living in a big house in a supposedly nice neighborhood hadn't solved a single one of my problems, and there were probably thousands upon thousands of people living in trailers who lived lives infinitely happier than my own. But the trailer shared by Gary and Jody was not that place. It was exactly what I thought of when I thought of a single-wide trailer in the middle of nowhere: a complete and utter dump, with brown wood paneling bowing off the walls, broken ceiling tiles, and trash everywhere. Empty beer cans and Mountain Dew bottles littered the floor. Filthy, used pans and baking sheets covered the kitchen countertop. A sink full of dishes. Food packaging everywhere. A giant flatscreen TV sitting on a pair of boxes. And not a single gun in sight.

I moved down the hall to the first bedroom, with a broken accordion door and a red sheet pinned over the passageway. I swept the sheet aside with my arm, hating to touch it. This, I realized, must be Jody's room. There were porn magazines on the floor, a fake samurai sword on the wall, a dirty mattress with no sheets, and a dozen bottles filled with dip spit. Dirty clothes everywhere, and a smell that somehow suggested both cum and diarrhea.

Gary's room was only slightly better, with a bunch of tin signs nailed to the wall, for Pabst Blue Ribbon and Coca-Cola and railway crossings. A dresser missing a drawer, and an unmade bed with unmatched sheets. Still no sign of any guns. The rug on the floor was so disgusting, I hated to do it, but I got on my belly to search under the bed.

After pulling out a bunch of trash, I would have breathed a sigh of relief if I hadn't been holding my breath. There were the guns. I pulled out a cheap AR-15, three old .22 rifles with wood stocks, and, wrapped in a dirty, blue sleeping bag at the very back, my grandfather's Remington 1100. I took the gun from the sleeping bag, replaced it with the one I'd brought, shoved the other guns back into place, and kicked the trash back under the bed.

The front of my shirt was covered in dog hair and dust, but I didn't take a moment to clean off. Now I began to run. I turned off the lights, locked the front door, and put the key back on the porch light. I put the shotgun into the case and tossed it in the backseat, then backed out and took off.

I had done it. The gun was mine again. I couldn't believe it.

As always, there was still a problem. The fucking dog. I couldn't leave him out there.

I drove in the direction I thought he had gone. There wasn't a car in sight, so I went slowly, looking both ways, weaving across the highway to shine my lights into the brushy roadside. I couldn't find the little bastard. I was about to turn around when I almost ran it over. It shot across the road right in front of my car like it was being chased. Hell, it probably ***was*** being chased.

I slammed on the brakes and wheeled the Tahoe around to shine my lights in the direction the dog had gone. I rolled down my window and started yelling, "Hey, man! Hey, man! Come here, puppy." Then I saw him. He was standing at the base of a little mesquite tree shivering about ten yards off the road. My voice is what stopped him, so I kept talking. "Hey, little man. Come here, little guy. You're okay. You're okay."

I put the Tahoe in park, got out and started walking over to him, talking the whole time. I was sure he'd run away, but as I got closer, he got lower and lower to the ground, shaking all over until he was curled into a trembling little ball by the time I got to him. I had just reached him when a car horn from the road almost sent him running, but I grabbed him as he stood to take off. A pickup had almost hit my car, parked halfway on the road with the driver's door still open.

I got back into the Tahoe with the still-shaking dog in my lap and headed toward the trailer. But as I drew near, I saw that the pickup—a ragged F-150 with peeling paint—belonged to my cousin Jody, who, fatter than ever, reached for the key and put it in the lock.

By the time I got to the main highway, the dog was squirming around licking my hand. He was pretty fucking cute, and the more I talked the more he seemed to like me.

When I got home, I hated to wake him. He'd slept curled up in my lap for hours.

I didn't have a leash, but I took a risk and set him down in the yard, where he promptly peed and pooped and followed me inside with my grandfather's Remington 1100 slung over my shoulder. Inside, he sniffed around and then took off running in every direction, slipping on the wood floor.

I took out the Remington and placed it on my grandfather's table. It was a reunion for the two of them. They'd been together plenty of times before. I sat on the couch and looked at the shotgun, amazed I'd actually done it. The dog jumped up beside me and, after turning in a circle, curled up for a nap. I scratched his head and looked up at the clock. It wasn't even midnight. Fuck's sake, it was still Saturday.

# 47

I WOKE UP to take the dog out at ten, fed him some scraps, and then went back to bed. Later, I DoorDashed lunch for myself and dog food for him. When it arrived, he went wild with excitement. He went out with me for my cigarettes, never wandering far.

I wondered what I would do with him. Maybe Tonya or Luis would take him or know someone who could. The weird part was that when I spoke to my mother, she updated me on everything imaginable with regard to Jody and Gary, but never once mentioned the missing dog. I was hoping she would at least bring it up so I could find out his name.

Tonya came over that evening and immediately went crazy over him, as I figured she would. The feeling was mutual. He was all over her.

I told her the whole story about my uncle's stroke, and she got big points for remembering that he was "the asshole who stole your grandfather's gun." By the time I finished, she was speechless. She shook her head at me like she couldn't believe it. She looked at the gun on the table.

"So, that's the real one?"

"That's the real one," I said.

"Holy shit," she said. "You always told me you were crazy, but I guess I didn't really believe you were that crazy. And you're a dognapper to boot."

"Yeah, well, that definitely wasn't part of the plan."

"What's his name?"

"No idea. I've been calling him 'dog.' But he answers to 'hey, man.'"

"Well, that's not a dog's name. How about Remington, in honor of the gun?"

"That works," I said.

"Remmy?"

"Yeah, Remmy. That's not bad."

The dog wagged his tail and looked up at us.

Remmy was a happy diversion. I even took him over to Luis's house. Remmy liked the cats a lot more than they liked him, but once he calmed down, he sat beside me, listening to us talk.

The next two days were filled with little chores related to Remmy, including a trip to the pet store to get him bathed and to have his nails trimmed. The girls who worked there fawned over him and told me how well he behaved. While we were there, I picked up some dog food, a bed, and toys for him to play with until I figured out what to do with him.

Ruck called on Monday night, so I got to tell the whole story a third time. I didn't embellish a word, but I could tell he was impressed.

"Well, that's totally fucking crazy," he said.

I texted him a picture of Remmy.

"Look at the picture I sent you. He's pretty cool, right?"

"Yeah, looks to be."

"I always thought chihuahuas were kind of yappy and annoying, but he's not like that at all. He's real mellow most of the time. He sleeps under the covers right beside me. Never had a dog do that before. He's a trip."

"You get any shooting done?"

"Yeah, I did. I went out to my buddy's land. Went pretty smooth. I'm not as bad a marksman as I thought."

"Glad to hear it. So, you're ready, then?"

Ready? I could hardly think of what the word even meant. Of course I wasn't ready. But that didn't mean I wouldn't do it.

On Tuesday, after I went to the rental car place, I talked to Sam, who was already deep into planning the wedding. I kept wondering if I should find some way to say something profound like someone in a movie would, about how much I loved her, and if anything happened to me, etcetera, etcetera. But it seemed like bad mojo talking that way before a big project. Still, I would write letters to her and my mom in case anything happened. Another thing to add to the list.

After we hung up, I sat on the couch, petting the dog, half watching *Heat* for the thirtieth time, recounting to myself the story of my trip to Houston. I wished I could've told Sam and my parents. Maybe years down the road, I would confess. "Remember that time Gary had a stroke? Well, I told you I'd end up with grandad's Remington, didn't I?"

When I fell asleep that night, with Remmy at my side, I dreamed the junkyard was an enormous pirate ship. Under cover of night, I pulled alongside in my little rowboat and hopped aboard. With no sounds but the wind in the sails and the water lapping against the ship, I went from cabin to cabin, slashing the throats of the pirates one by one, stealing away with their treasure.

# 48

I WOKE UP knowing it was the day, the same way I'd woken up on the days of my weddings and divorces knowing that something big was happening. When Remmy could sense I was keyed up, he stood on my lap licking my face. I was already wondering if I should keep him.

In some ways, it was a day like any other: I did bumps off the kitchen counter and DoorDashed Chick-fil-A. I dodged phone calls from my mother and had *Extraction* on the TV as I sat at the breakfast table loading magazines. I spent an hour looking for a couple of Glock holsters. When I found them, I dropped them and the Glocks into an oversized duffle bag along with the 300 Blackout AR and a bulletproof vest. I put the bag next to the front door and kept thinking about what else I should put in there, adding things throughout the day. A ski mask. A flashlight. Binoculars. No such thing as being overprepared.

When Hector texted to remind me when and where to meet him—Luis's house at six o'clock—I replied right away, reassuring him I would be there.

The day passed. When the sun hit the TV, obscuring the closed captions, I didn't take a nap like I normally would have.

I didn't have to remind myself why I was doing what I was doing. The housekeepers, the rent, the wedding. I still needed a new washing machine. I'd all but run out of clean clothes, sniffing my T-shirts before I put them on. I was tired of thinking about all the things in my house I could sell if push came to shove. The litany always turned into a list of all the things I would never think of getting rid of. Never that guitar. Never that motorcycle. Never those guns. I dreamed of being a minimalist like Ben Affleck in *The Accountant*, but I never knew how to do it when I had so much stuff that was worth keeping.

At six o'clock, Tonya arrived. Remmy was all over her. She was already his favorite person aside from me.

"What am I not allowed to clean or organize while you're gone?" she said.

"Why don't you chill out and watch a movie? That's what I'd do."

"When are you going to be home?"

"Might be late. If I'm not back by two, there's some stuff in my bedside table I need you to handle, but only open it if I'm not back by then, okay?"

"You'll be that late?"

"Probably not. I'll probably be back before midnight."

"How worried should I be?"

I ignored the question. "The dog eats at seven. His food is on the counter. Give him like a third of a can. And he likes fresh water with his dinner."

I lifted the bag and hoisted the strap over my shoulder.

Tonya watched me. "I thought this wasn't a gun thing."

"I'll tell you about it later."

I kissed her on the cheek and left.

Before I turned off my street, Hector texted. *Hey man, you're late.*

I ignored the text and drove to a parking garage at a mall near my house, where I parked the Tahoe, left the keys on the front tire, collected my things, and got into the black Ford Escape I'd rented the day before.

I texted Hector. *Nearly there. Having car trouble. Give me a minute.*

*K.*

It was 6:20. I was supposed to pick him up at six.

He sent three more texts over the next twenty minutes. He was getting really tense by then.

*Sorry, man*, I texted. *I've almost got this straightened out. I'm close.*

The texts kept coming until seven o'clock, by which time we should have been on the road a half hour earlier.

I sent him a text that said, *Broke down again. Stuck with this cop on the side of the road. He's trying to help, and I can't get away. I'm really sorry. Catch up with you tomorrow. I can give you a ride then if you can postpone.*

Hector, who I'm sure was pretty pissed, did not reply.

Even if he found someone else to go with him or went alone, which I didn't think he would do, he'd be late.

The whole ride, I tried not to think about anything. You don't worry when you're driving to the office, except maybe to dread the workday. You're just going to do a job.

When I was ten minutes away, I pulled alongside some dumpsters at a gas station and arranged things. Holster at six o'clock, holster at two o'clock. I put extra magazines for the Glocks and the Blackout in the pockets of my jacket, remembering where they were, like I'd decided to do days earlier. These in this pocket, these in the other.

I got the Blackout out of the bag and set it on the seat beside me. I did a bump off my thumb joint and lit a cigarette. I rolled the window down and could smell the dumpster. It was a chilly night.

I put the car in drive.

When I got to the junkyard, I pulled right in. It was eight o'clock, the time they'd scheduled for Hector, though I was sure he'd rescheduled. It was the same as the time before: two guys on the porch. One of them sitting on the railing talking to the other one who was standing up, smoking a cigarette. Maybe they were the same guys I'd met the last time. I couldn't tell. They were nothing but silhouettes.

I put the car in park like I was going to wait for the guy to come to my window again. The guy sitting on the railing still had his back to me, but he was turning his head looking over his shoulder like "what does this asshole want?" He was about to make a show of getting up real slow like I was a pain in his ass. But that was fine. I reached for the Blackout, aimed at him before he could respond, and shot him in the center of his back, blasting out his lungs. He fell forward with enough force that he slammed into his compatriot's legs, which slowed the second guard as he tried to raise his AR. My aim was on him like I hadn't even put it there. His head snapped back and didn't snap forward again.

The gun was extremely quiet. It seemed strange I hadn't spent more time shooting Blackout. Maybe it would be the next thing I got into.

I didn't turn off the engine. "Hysteria" by Def Leppard was starting on the radio, which I took as a good sign. I put on the ski mask and a pair of black leather gloves, got out and walked up to the porch. The two guys were on the ground to the left of the door, and I only glanced to see if any movement was coming from their direction. I didn't want to look in any detail. I did notice their guns, though. They'd retired the ridiculous, colorful guns they had before. These were black AR-15s, like the ones I had built.

There was a decision to make in this moment that I'd been delaying. Did I do something with the two guys on the porch to move them out of view of passing traffic or whoever might show up for their next appointment? Or did I leave them and hurry the fuck up? I thought about their guns. I wanted them. Something that always bugged me in movies was when people left perfectly good guns behind. This was the problem with planning and not planning. You didn't know what all you had planned and what all you hadn't planned.

One of the guns was almost free of the man's grip.

Hurry the fuck up, man.

I turned the knob on the little building and went inside. I thought maybe it would be a real sophisticated operation with monitors showing surveillance footage from around the place, but it looked like a regular junkyard office. Desks with computers and filing cabinets, stupid posters on the walls. I tried switches until I found the one that turned off the porch light. I searched for the gate control, expecting a big red switch like in an action movie, but it was just a little doorbell button.

I stood there a moment. Fuck it. I pushed the button, and the gate began to move.

I stepped down off the porch and walked quickly to the gate. I raised the gun as it opened to reveal the man on the other side. He was startlingly close, less than ten feet from me, not even touching his gun. He was looking down lighting a cigarette. I pulled the trigger, and he fell. It was an extremely mundane looking death. It wasn't a movie. It was like he was there on his feet and then he wasn't on his feet. He wasn't a person at all. He was a neutralized threat.

Now here was another problem. His body was in the middle of the lit driveway. His face was now a scramble, and I didn't relish the idea of dragging him. My knee was already bothering me, and it would only get worse the more I used it. Fuck it. I grabbed his arm and pulled the guy, leaving a trail in the dirt. Another black AR. This time I did grab it. I went back to the car, tossed his gun in the backseat and got in. "Hysteria" by Def Leppard was ending.

I didn't like the shape of the junkyard. This weird section you had to drive through to meet with the dealers was kind of eerie. I watched myself from above, creeping along with my headlights off. I had a decision to make about what to do next. I could drive up and try to do what I'd already done with the first two guards, but I figured the guys at the back of the junkyard were going to be a lot more suspicious of a random car showing up, which would mean drawn guns from the moment I rounded the corner. But I really did not feel like walking.

Another problem: the path was too narrow to turn around in, so even if I did take off walking, once I got back to the car, I'd either have to reverse out of the junkyard or drive up to the single-wide to turn around. It was no time to be lazy. Once I was pretty sure I was close to the last turn before reaching the trailer, I stopped the car. "Stranglehold" was on the radio. That seemed like a good sign. I sometimes felt that no one was talking about Ted Nugent anymore, which seemed pretty fucked up, actually.

I did a quick bump, then got out and started walking without closing the door to the Escape. After a minute, it was already the most walking I'd done in months. My knee hurt. I'd have to go to the doctor soon, sitting in fucking waiting rooms and getting referred around. It was frustrating even to think about it. I had the Blackout at my shoulder, scanning the terrain before me as I approached, feeling, I admit, extremely professional.

I got to the last set of junked cars before the single-wide, took a breath, wished I could light a smoke, and peeked around the corner. There was a man sitting on the steps leading up to the single-wide. He was looking at his phone. This whole generation was looking at their phones all the time. The thing that bugged me was that the last time there had been two guys and now there was only one. Did that mean the second guy had taken the night off, or was he taking a leak behind a car, or was he inside the trailer bullshitting with the other guys? This was what was frustrating. You didn't fucking know.

I waited a couple of minutes, and the guy never looked up from his phone. I figured two minutes was enough time for most guys to piss, so I went ahead and shot the guy on the porch. That silencer was so quiet, it blew my mind. The guy slumped forward and fell off the step. The loudest sound was his AR clattering to the ground, and I doubted it was loud enough for anyone inside to notice.

I started walking up to the trailer. It looked a lot like my uncle's, though it was in a hell of a lot better shape. I was pretty much out in the open. I could have sort of flanked to one side, maybe, stayed in the shadows a bit more, but it would have taken twice as long and twice as much pain in my knee. Plus, I was sure there was no one there. It was just me and a dead Mexican.

A sound rang out, and I heard a bullet hit a car behind me. It was so loud, I sort of zoned out for a second. I looked up and there was the other guy, who I guess took longer to piss than I expected. He was ten yards away. The only thing I couldn't understand was why he had stopped shooting. When I spotted him, he was doing something to his gun, fucking with the selector switch. The fucking thing had malfunctioned. I couldn't believe my luck. I raised my gun and fired. I guess there was something going around, because I missed him, too. He started to turn to run, but I recovered and put two bullets in him pretty tidily.

The guys inside the trailer had definitely heard the sound outside by this time—from their buddy's AR, not my Blackout.

If I had to make an educated guess about what they were doing in there, I figured it was safe to assume they were: 1) calling for backup, and 2) preparing to get into a gunfight with whoever was outside. This time, I did not worry about my knee. I jogged as quickly as I could, heading to the side of the single-wide, away from the door.

I was wrong about what was going on inside, because the door began to open, and the man who opened it said, "Hey, motherfuckers..." before noticing the body at the bottom of the steps. I surmised in those two words that the fellas had been reprimanded more than once for firing weapons, and this guy was sick of telling them to knock it off. The way I was standing, the door blocked me from seeing much of the guy. I fired twice through the door, and he fell forward, propping it open. His gun landed beside him.

There was a long pause in silence. If I was right, only two living people were left in the junkyard.

"I'm not armed," a Mexican voice called out.

"Oh yeah?" I said. "I am."

"I can throw the coke and the money out the door. You don't have to come inside."

"Yeah, I bet you'll hand it all over without keeping any for yourself."

"Well, how can we do this where one of us doesn't get killed?"

"I don't mind if one of us gets killed," I said. "Because that one of us is not going to be me."

"What if I come out with my hands up?"

"That would be good for you."

"Okay, I'm going to. If I do, you won't shoot me, right?"

"I don't remember saying that, but we can go with that for now. If you come out with a weapon anywhere on your person, I will shoot you in the head."

"Got it," he said.

The guy walked out the door with his hands in the air. He was a short, little pudgy dude in cowboy clothes. I felt like I got a pretty good sense of the guy in a single glance, like he was probably kind of a cutup. A class clown type. I figured they'd have some real ruthless guy in there, but this guy didn't look the part at all.

He looked over at me in my ski mask and said, "Hey, Batman."

"Hey."

"We got a leak somewhere or what?"

"A leak?"

"You're here now because of a special reason, no?"

I assumed he was referring to Hector being late for his appointment. It was exactly what I had worried about, that somehow the raid and Hector would be linked. "What's so special about tonight?"

"I think you know."

"Keep stalling, asshole, and see what happens."

"The shipment. Tonight we got our biggest shipment. Kind of funny you'd come tonight without knowing that."

"Maybe I just got lucky."

It's true that, once I got on a lucky streak, it seemed like there was no stopping it.

"Can I put my hands down?"

"No. I should shoot you and get out of here."

"I could help you load it. Then you could let me go. I won't tell anyone about you. I haven't seen your face. I don't know who would have hired an old white guy, and I have no idea how I would figure it out."

"How much coke is there?"

"Sixty kilos."

Because I had spent a lifetime pretending not to be surprised when people said surprising things, I hardly blinked.

"How much cash?"

"Cash? Maybe thirty thousand. Something like that."

"That's it?"

"Sixty kilos and thirty grand isn't enough? What the fuck did you think you were getting?"

He had a point.

"Is there anyone else on this property? Aside from the three guys at the front and these three here?"

"I only see two here."

"There's another dead guy over there," I said.

The pudgy Mexican craned his neck and said, "Oh, yeah, I see that. That's everybody. I thought we were sort of overstaffed, but I guess you proved me wrong. Where's your car?"

"Around that corner."

"Cool, go get it, and we'll load your shit up, man."

Fuck, what was I supposed to do with this dude? I had no fucking idea. I looked at my watch. "I need to hurry," I said. "I want you to take me inside and show me the stuff. But I'm telling you now, if there's anyone inside, and you don't warn me now, I'm going to shoot you first, even if it's the last thing I do."

"There's no one inside, man. Like I said, this whole thing has really revealed some flaws in our security."

I put the gun to his back, and we went inside. It was almost the exact same layout as my uncle's place. It smelled like weed in there. There was a table in the living room where they presumably did business and a pair of recliners in front of a TV with a video game system hooked up to it. The first room, which would have been Jody's bedroom, had another table in it. That table held a pile of sixty bricks of cocaine.

"Where's the cash?"

"There's a bag in the closet."

"Show me."

The pudgy Mexican went to the closet, opened the door and took out a bulging satchel.

"That's not thirty grand," I said. "Unless it's in ten-dollar bills."

"It's sixty grand," he said. "But you were expecting thirty, so you should be happy with thirty."

"Give me a fucking break," I said. I put out my hand, and he gave me the bag.

"How are we going to get all this coke into my car so you don't get shot? That's the thing I would be worrying about if I were you. Not trying to fucking bilk me out of thirty grand."

"Are you going to shoot me either way? Because I'm not going to help you load your car if you're going to shoot me."

"I won't shoot you if you figure out how to load my car in such a way that I don't have to."

"There are garbage bags under the sink, man. You can bag up all the coke, and I'll help you carry it."

"But then you'll see my car and come after me."

"I guess you could blindfold me."

"This is getting too complicated," I said.

The truth is I didn't want to shoot him. I'd barely seen the faces of any of the other guys I'd shot. The closest I'd gotten was the dude with the fucked-up AR, but he was mostly looking at the gun, not me.

It took a few minutes of work, but soon we were walking towards my car with half a roll of paper towels wrapped around the guy's head. He was carrying four garbage bags full of coke. I had two, along with the bag of money and the 300 Blackout.

"How much are you paying for ARs?" I said when we walked past the body of the guy with the fucked-up gun.

"Ten grand."

"Seriously? You can get an AR for like a thousand dollars. Probably less."

"Yeah, but these are full-auto."

"You're shittin me. Where the fuck are you getting full-auto ARs? Russia or some shit?"

"We have our sources."

"I want the guns, too," I said.

"You want the guns?"

"Yeah, all of them."

"Well, I can't carry the coke and all the guns, too."

"We'll come back for them."

That's what we did. We loaded the coke and cash into the back of the Escape, then grabbed the guns and put them in, too. It was a real pain in the ass, considering he couldn't see what he was doing.

I should have at least walked the guy back and tied him up or something, but I was too burned out, and my knee was killing me.

"Give me your phone," I said.

"Come on, man. Don't take my phone."

"I'll put it up front."

"Up front where?"

"On the porch to the building. You're lucky I don't shoot you. Walk back to the trailer and play a video game or something. You can come out in half an hour."

The guy sighed and handed me his phone. "Okay, man. *Buena suerte*."

"Yeah, you too."

I watched him walk back in the direction of the trailer, then got into the car and started reversing to the entrance. The car had a backup camera, so it wasn't that bad, actually.

When I got to the front, I did something that someone else might think was pretty dumb. I stopped the car in front of the building entrance and went up the steps. I took the guy's phone out of my pocket, set it on the rail, and then bent down to retrieve the guns off the two guys there. They were mine, I now realized. I had built them, and I wanted them back. The Escape was parked in front, engine on, headlights on, driver's side door open.

It was then that I saw a pair of headlights. I watched from the darkness of the porch, where I couldn't be seen, as Luis's white pickup pulled into the driveway. I could see their shadows inside, Luis at the wheel and Hector beside him, the two of them hesitating at the irregularity of the scene. I stayed frozen, but I saw they weren't going anywhere. Luis flashed his lights. When nothing happened, he tapped the horn, making a weak little honk.

I pulled the AR off the first guy I'd shot, feeling his dead weight slip out of the strap, took aim and shot out Luis's driver's side mirror. After handling the silenced Blackout all night, the sound was unbelievably loud. My ears began to ring instantly. More hearing damage. Just what I needed.

Luis threw the truck into reverse with admirable speed for an old guy. Then, with a squeal of his tires, the two of them were gone. The other AR was too covered in blood to take with me, so I settled for the one in my hands. I went down the steps, tossed it in the back, and left the way I came.

# 49

I DITCHED THE Escape where I'd left the Tahoe and drove home shaking. Not as bad as the night with Nikki and Diesel the Rottweiler, but bad enough. I was sweating. All I wanted was to smoke a joint, drink a bottle of wine, and take two Xanax. Tonya met me at the door with Remmy in her arms.

"You made it!" she said.

"Of course I made it."

"You look worn out."

"I am."

"Come sit down."

"I will. But first could you give me a hand getting sixty kilos of coke out of the Tahoe?"

The house, when I finally got around to noticing it, looked better than it had in ages. Tonya hadn't had enough time to put that much of a dent in the insanity, but she'd broken down a million Amazon boxes, cleared a bunch of trash from the coffee table, and organized the piles of unopened mail into rows of neat stacks. My mother wouldn't have been impressed, but at least it didn't look like a complete loser lived there.

Tonya broke up the weed for a joint, and we smoked it on what was left of the deck while I told the parts of the story I thought she could handle. Sometimes it felt like life wasn't worth living if I couldn't tell someone about it. She listened and seemed mostly alright with it.

After she left, I called Ruck, who had stayed up waiting for my call. He sounded so worried, I felt like shit for not having checked in sooner. I got to tell the story again, this time with all the gory details.

"Jesus, Wince," he said. "For better or worse, you fucking did it, man."

The next day, Luis would tell me his story, too, about squaring off against a gunman outside the junkyard. And how it was all my fault, of course, because of my dumbass car troubles.

Of the five AR-15s I recovered, I sold four back to Hector after cleaning them up and testing them at his land. The one that had malfunctioned only needed a slight adjustment to make it right. For once, my laziness had paid off. I kept that one as a souvenir. Maybe one day I'd mount it on the wall next to my grandfather's Remington.

When I met up with Hector to sell him the guns, I heard his version of what had gone down at the junkyard that night. I couldn't tell if he looked at me with suspicion when he mentioned that some fat, old white guy had somehow robbed the cartel, though he did mention more than once they were moving operations, and the next guy who tried that shit was going to be in serious trouble. It was lucky for him, he said, that he himself had been fired upon by the robber. Otherwise, they would have thought he was involved.

Getting rid of that much coke would be a full-time job, and I didn't feel up to figuring it out right away. The cash meant I could get a new washer and pay six months of rent in advance. I had enough to pay for Sam's wedding and chill the fuck out at least for a while.

I drew out a plan for the deck and hired the guy who'd built my neighbor's. Who was I kidding thinking I was going to do it myself? The crew made fast work of it, and when it was done, it didn't look as good as what I would have built, but it was good enough, and there was a set of steps for Remmy to use so he could run around in the yard while I smoked.

Once the deck was finished, I asked Tonya if she felt like housesitting while I went on a trip. Well, housesitting and dog sitting.

"It'll clear your debt to me if you do," I told her. "The rent is paid, but I might be gone for a few months. I think I'll see how I like Indonesia."

It was a good time to get away from the coke, and I figured she could probably use a place to be. Plus, it might be my last chance to see if sleeping in a hut on the beach was everything I'd imagined.

It used to really bug me when I thought about how a younger version of myself would see me now. How he'd laugh his ass off at a broken-down old man in flip-flops and cargo shorts. Who hadn't ridden a motorcycle in years. Who maybe didn't have all his shit perfectly in order. But the thing that younger version of myself didn't realize is that I'm laughing my ass off at him, too. He might be a badass, but he sure as fuck doesn't know what he doesn't know.

"The man, the myth, the legend, Winston Fisher, Jr.," my second wife, the whore, used to call me in a pretty bitchy tone. It was her shorthand way of accusing me of being an arrogant, self-aggrandizing asshole. It bothered me when she said it, but now I'm glad she did, because I had to think up a way to defend myself. Ultimately, the thing I landed on was that if I don't make a myth of myself, no one else will do it for me.

# ACKNOWLEDGEMENTS

Special thanks to editor
Jess Bugg

to illustrator
Griffin DeNigris

and to publisher
Chris Stoddard

## ABOUT THE AUTHOR

Author Photo by Travis Willman © 2026

DREW NELLINS SMITH is the author of *Arcade* (Unnamed Press, 2016). His nonfiction has appeared in *The Los Angeles Times*, *The Washington Post*, *The Believer*, *Tin House*, *VICE*, and *Literary Hub*, among others. He lives in Austin, Texas.

# BOOKS BY ITNA

*The Dark*
Leonid Andreyev

*Aaron's Rod*
D.H. Lawrence

*The Beads*
David McConnell

*@UGMan*
Mark Sarvas

*The Virtuous Ones*
Christopher Stoddard

www.ingramcontent.com/pod-product-compliance
Lightning Source LLC
LaVergne TN
LVHW031334150826
845673LV00012B/2889

* 9 7 9 8 9 9 8 9 5 5 1 2 9 *